BIOLOGICAL SYSTEMATICS

HERBERT H. ROSS
University of Georgia

ADDISON-WESLEY PUBLISHING COMPANY, INC.
Reading, Massachusetts • Menlo Park, California • London • Don Mills, Ontario

This book is in the Addison-Wesley Series in Biology

Johns Hopkins, III
Consulting Editor

Preface

In the preparation of this book I have had three primary objectives. The first is to emphasize the scientific aspects of systematics, because only when this is done can many areas of argument be put into their proper perspective. The second is to use examples from animals, plants, and bacteria whenever possible. This integrated approach is out of the conviction that, in systematics, botany, microbiology, and zoology can learn much from each other. After all, living things of every type have much in common. The third aim is to present the theory and practice of systematics simply and clearly, in the hope of offering students in any field a well-rounded introduction to the nature and promise of systematic research. Such a broad view can provide an understanding of the basic concepts of systematics essential not only for systematists, but for geneticists, biochemists, ecologists, biogeographers, physiologists, geologists, and investigators in other areas of science.

The literature pertinent to systematics is too voluminous to be cited here in its entirety. Whenever possible more recent reviews or summaries have been cited that contain references to earlier or more specialized literature. Inadvertently certain critical papers may have been omitted; I would appreciate having such omissions drawn to my attention. Space has permitted the citation of only key papers from the very large and relatively recent number of contributions containing arguments pro and con on various systematic procedures. The interested reader will find many of these in the journals *Systematic Zoology* and *Taxon*.

Athens, Georgia   H.H.R.
November 1973

Acknowledgments

Many people have been extremely helpful in the preparation of this book. I am indebted to the following for reading some or all chapters and for making valuable criticisms and suggestions: R. H. Crozier and S. B. Jones, Jr., University of Georgia; G. F. Edmunds, Jr., University of Utah; D. L. Hull and L. H. Throckmorton, University of Chicago; C. A. and J. P. Ross, Western Washington State College; P. H. Raven, Missouri Botanical Garden; P. D. Ashlock and C. D. Michener, University of Kansas; and E. I. Schlinger, University of California, Berkeley. I am especially indebted to Dr. Jones for providing many of the botanical examples. In the light of some conflicting advice, full responsibility for the views expressed is mine.

Especially helpful in providing counsel and illustrative material have been K. G. A. Hamilton, J. C. Morse, P. E. Thompson, and W. J. Wiebe, University of Georgia; Hans Tralau, Swedish Museum of Natural History; P. W. Smith, Illinois Natural History Survey; W. P. McCafferty, Purdue University; R. K. Benjamin, Rancho Santa Anna Botanic Gardens; J. C. Frye, Illinois Geological Survey; B. L. Turner, University of Texas; H. E. Kennedy, Biosciences Information Service; R. S. Cowan, National Museum of Natural History; and D. F. Hoffmeister, University of Illinois.

Others too numerous to mention individually have aided with discussion and information on many specific items. Over the years all my graduate students have been extremely helpful and instructive, their discussions providing vigorous stimulation.

Much of my own research that forms the basis for many ideas in this book was supported generously by the Illinois Natural History Survey and the National Science Foundation. I am happy to acknowledge this with deep thanks.

I am grateful to many publishers and authors for permission to reproduce a number of illustrations and to others who kindly loaned illustrations for use in this book. Specific acknowledgment is made under each one. To Prentice-Hall, Inc., I am grateful for permission to paraphrase parts of Chapter 7 in *A Synthesis of Evolutionary Theory*. Finally I wish to thank the staff of Addison-Wesley Publishing Company for the preparation of many original figures and for editing the manuscript.

                                                                    The Author

Contents

1

Systematics and Its Development

Systematics began as a relatively simple activity including only species recognition and classification, then evolved into a highly complex science. It is now a broad field embracing many diverse avenues of biological study united by having concepts of species, phylogeny, and classification as a common basic core.

Systematics has frequently been termed the science of species, but this is not sufficiently exclusive. Comparative anatomy, genetics, and other branches of biology also deal with species. Since any science can most accurately be defined in terms of its special concerns with its subject matter, it is appropriate to examine the special concerns of systematics regarding species, and how these concerns developed.

EARLY SYSTEMATICS

The beginnings of systematics were undoubtedly rooted in our most ancient history. When primeval man developed language, possibly two or three million years ago, he must have been a biological systematist of a primeval sort. He would not have had a language without something to talk about, and this something would have included items in his natural environment. He would not have gotten far in talking about this environment without systematizing individual phenomena into more general concepts. This mental activity involves classification. Perhaps the first systematic concepts had an ecological basis, such as *animals man eats* contrasted with *animals that eat man*, or *berries that are bitter* compared with *berries that taste good*. But we do know that by the time man had evolved tribal life and freely articulate language he had developed species concepts and had correlated ecological phenomena for hundreds of different plants and animals as well.

It has been a long journey from this primitive empirical systematics to the scientific and interpretive systematics possible today. The journey probably started more than a million years ago, but it had little chance of progressing until man had developed written language, which enabled him to record not only his own observations but also to know what others had seen. The early written language alone, however, contributed little to advancement in systematics. These early written expressions of languages, dating back to 3000 or more B.C., were used first for expressing and recording the everyday needs of commerce and the cities. The more sophisticated urban peoples appear either to have lost their ancestral interest in natural phenomena or to have combined it with religious associations and mysticism.

The first steps toward modern systematics came in the sixth and seventh centuries B.C. with the birth of objective scientific investigation and deduction in the Ionian city of Miletus. Thales, the first scientific proponent, was primarily an astronomer; his student Anaximander introduced an evolutionary idea for the origin of living things. A later student, Anaxagoras, carried this objective type of thinking into the fifth century B.C. (Durant, 1939).

The first serious beginnings of biological systematics were exemplified by the studies of Aristotle about 300 B.C. Greek scientists of that period developed a remarkably objective view of all natural history, greatly improved Anaxagoras' concept of organic evolution, and developed a workable classification for all the life that could be seen without a microscope. But the subjugation of the Greek world by Rome in the first century B.C. seems to have stopped completely this progressive scientific spirit. The remarkable Greek discoveries in anatomy and physiology came to a halt, not to be rediscovered for 1500 years. In the religious and political turmoil of this long period, superstition and myth crept more and more into systematics and the objective ideals of Greek scientific thought were forgotten. One may marvel at how the Greeks learned so much. Even more astonishing is the fact that this knowledge and the inquiring spirit which motivated its discovery disappeared so completely for 15 centuries.

The exploring voyages undertaken by western Europeans of the 13th and 14th centuries appear to have signaled a second step toward modern systematics through the renewal of adventuresome thinking in western civilization. Remarkable discoveries in geography undermined the authoritarian dependence on classic writings during the Dark Ages of Europe. In biology the first firm steps out of this mental morass were made by a handful of isolated, highly discerning, and courageous investigators. Vesalius' work on human anatomy (1543) rejuvenated observation, and Harvey's demonstration of the arterial and venous circulation of the blood (1628) reintroduced experiment, virtually dead since the third century B.C.

Contemporaneously with Vesalius' and Harvey's advances in anatomy and physiology, new concepts arose in systematics. During the Dark Ages the literature on systematics comprised books on natural history and herbals that dealt with economic and medicinal plants and their uses. Most of this material was copied from previous works. Fuchs (1501–1566), a professor of medicine at Tubingen, Germany, grew many of his medicinal plants and observed their characteristics. This was the beginning of more accurate observation and helped to rekindle curiosity about plants. In 1583, the Italian Cesalpino, in *De plantis libri*. stated with reference to species that like always produces like, the precursor of a biological species concept. The Englishman John Ray, working with both plants and animals, reiterated this same concept of a species a century later.

During the seventeenth century the tempo of biological discovery quickened. Introduction of the microscope led to the microanatomical works of Malpighi and Swammerdam, and to the discovery of microorganisms with which Leeuwenhoek astonished the scientific world. Museums of ethnology and natural history sprang up in many European cities, rivaling Aristotle's long-lost collections in his Lyceum with regard to regional species, and enriched by specimens brought from every corner of the globe by explorers and travelers.

Systematics blossomed in the eighteenth century, culminating in the remarkably comprehensive classifications of the Swedish systematist Caroli Linnaeus and his followers.

At this time the origin of each species was considered to be a separate act of divine creation. Species were regarded as immutable or unchanging, continuing generation after generation in the same mold. This attitude may seem somewhat surprising because through historic time man himself has changed perceptibly from generation to generation. Moreover, domestication and trait selection for various desirable characteristics produced changes in the cattle, dogs, and other animals man added to his civilized retinue. But man has been loathe over the years to consider himself or his artifacts as a basis for scientific thinking, and, except for the unpopular and soon forgotten idea of Empedocles (circa 450 B.C.) that man evolved from the apes, these perspectives went unnoticed. The period when species were thought to be immutable may be defined as the *no time, no change* era of systematics.

Despite their lack of a time-dimension or evolutionary outlook on life, the 18th century systematists made tremendous strides in their systematic concepts. Linnaeus was responsible for two major accomplishments: (1) a character-based classification of both plants and animals, giving a basis for the arrangement of specimens in collections and (2) the popularization of the currently used binomial system of nomenclature, useful as the basis of our information storage and retrieval system for the great bulk of biological information.

Linnaeus and his contemporaries (Haartman, 1751, 1764; Kölreuter, 1761–1766) were not only classifiers but experimentalists (Roberts, 1929).

They hybridized literally hundreds of species and lesser populations, and arrived at the conclusion that plants which did not hybridize successfully were species, whereas those which did were varieties of the same species. In addition, a number of pre-evolution systematists (both before and after Linnaeus) developed the idea of morphological relationships between supraspecific groupings such as genera and higher categories. For example, Linnaeus classified the plants on the basis of their floral parts, first grouping the species by the number of stamens each possessed. His student J. C. Fabricius classified the insects on the basis of their different types of mouthparts.

The two ideas, the existence of varieties and morphological relationships, undoubtedly raised questions that led to early concepts of what we now call modern or evolutionary systematics.

MODERN SYSTEMATICS

The next era in systematics, which can be thought of as the birth of modern systematics, accompanied the scientific revolution that swept biology during the late 18th and early 19th centuries. Up to this point, biological science had been purely descriptive with little concept of fundamental laws. Now these came to light in rapid succession. Lamarck and Cuvier established the idea of comparative anatomy; Owen advanced the idea of homology and analogy of parts. Milne Edwards propounded the idea of division of physiological labor; Müller demonstrated the relation between anatomy and physiology. Schwann and Schleiden demonstrated the cell theory. Bichat founded histology, Von Baer founded embryology, Schultze defined protoplasm.

That these discoveries were made in such rapid succession is not strange. Scientists had been on the verge of seeing them for years and as soon as one fundamental was discovered, it served as a key to unlock the next half-anticipated secret.

Contemporaneously an equally revolutionary set of ideas arose concerning species. Brought forth effectively first by Lamarck in 1809 (see Lamarck, 1963), these new ideas proposed that species were not immutable but changed with time, with the result that ancestral forms gave rise to various derived forms. Lamarck began his career as a botanist, then later switched to investigations in animal morphology and classification. The series of annectent changes he observed in various groups of animals apparently led to his conclusions concerning evolution by character change, a delightful example of a hypothesis reached by inductive reasoning. When he attempted a deductive set of consequences, he proposed that the changes were the result of the inheritance of acquired characters. On this score he had little data. His rival compatriot Cuvier seized on this weakness. He cut off the tails of mice for many generations, but the ensuing generations of mice continued to have tails. On this evidence, Cuvier ridiculed Lamarck and the entire idea of evolution into oblivion for half a century. Cuvier, however, could not stem the tide of inquiry forever, and many factors led eventually to the vindication of Lamarck's hypothesis. Important in this respect were the developments in geological investigation.

The discoveries of the 18th and 19th century geologists added a time dimension previously lacking. Through their studies it was evident that strange animals and plants no longer in existence had lived eons before and were now extinct, and furthermore that many forms living today were not represented in the faunal and floral remains uncovered in strata representing bygone ages. These diverse viewpoints and collections of data plus much information and thought of his own led Darwin to formulate his theory of natural selection as the cause-and-effect relationship between organisms and environment that brought about changes in lineages with time. Wallace hit on the same idea, which was presented jointly in 1858 as the Darwin-Wallace theory of evolution through natural selection. In 1859, Darwin published a summary of his ideas in the first edition of his *The Origin of Species through Natural Selection*. The Darwin-Wallace theory of natural selection provided a highly probable mechanism for evolution that Lamarck lacked and that countered Cuvier's ridicule of both Lamarck and the concept of evolution.

The application of this idea to systematics should have been explosive. Rather than being endless generations of individuals struck from the same mold, species were dynamic populations changing with time. Their present characteristics were undoubtedly different from what their future ones would be and definitely different from their past ones. In short, from the evolutionary viewpoint species had a history and changed with time. This understanding offered a new conceptual framework for systematics and set the stage for its development into many dynamic areas of study.

However, the development in systematics was not rapid because Darwinism itself encountered several decades of opposition. Immediately after publication of the *Origin*, religious leaders and many eminent scientists (including systematists) hotly contested the whole idea of evolution, often on a highly emotional basis (Mayr, 1972). Advocates of the idea, meanwhile, amassed more and more data supporting evolutionary theory, and by the turn of the century Darwin's ideas had won widespread acceptance.

At this time (1900) Mendel's work in genetic inheritance was rediscovered and extended. The character differences found in these early genetic studies were quite large compared with the minute variational character changes theorized by Darwin as the basis of evolutionary selection. Controversy immediately engulfed the idea of Darwinian selection, the geneticists claiming that the newly-discovered large genetic units must be the basis of natural selection.

As the intricasies of inheritance gradually became better understood, genetics and Darwinism were intermeshed. The resulting modified theories of natural selection were called *neodarwinism.*

The Species

As a result both of the holdover of the "immutability" concept and the lack of understanding of genetics, for some time after Darwin species were defined on the basis of the possession of distinctive characters. According to this concept, species were delineated by *character discontinuity.* If in a group of flies, for example, some individuals had only two hairs on the head and others had four, those with two head hairs would have been considered one species and those with four another. The same conclusions were drawn concerning plants with red or yellow flowers, mammals with black or red fur, or bacteria able to grow with or without a specific dietary ingredient. When geneticists began rearing and crossing individuals of plants and animals in the laboratory, they discovered that crosses between apparently identical individuals would sometimes produce viable offspring and sometimes not. The investigators were obviously dealing with two perfectly good species, each reproducing within its own kind but not with each other, and yet the investigators could not distinguish them on visible characteristics. As more examples of these cryptic or hidden species were discovered, the morphological discontinuity definition of species had to be abandoned. The concept of *genetic incompatibility* was substituted in its place as the criterion for separating species. At first this concept was widely accepted but it soon ran into difficulties. Botanical studies demonstrated that in plants:

1. Occasional hybrids occur between moderately distantly related species, and these hybrids may be able to backcross with one or both parents.

2. In local areas two otherwise distinctive species produce hybrid swarms that persist for long periods of time.

These findings undermined the concept of genetic incompatibility as a criterion of species definition. Zoologists long claimed that animal species did not hybridize in nature, but gradually examples of successful interspecific crossings between species began to accumulate. The arguments about species then resolved into the question: How much hybridization is possible between species for them to be considered species?

Attempting to reconcile these and other arguments about species, in 1951 Simpson proposed an *evolutionary definition* of species. According to his definition, species represent lineages that are evolutionarily independent in their passage through time. If hybridization occurred between them, it would not be frequent, effectual, or persistent enough to promote a fusion of any two lineages into one.

This definition is highly theoretical. If we could live another 30,000 or 100,000 years, we could watch two doubtful species and see if they became more distinctive with time or gradually merged into a single species. It is highly unlikely that human life spans will reach this length soon, and in any event we might want to get the answer now. To do so, it is necessary to devise operational techniques that will give us as good an answer as our data will permit.

These operational techniques take into account different mechanisms of species formation, genetic and evolutionary theory, the geological history and ecological changes of the areas and times involved, and any other factors that might be pertinent to the problem. We are constantly searching for new data and better ways of examining these data to achieve a better understanding of species, past and present.

When we think of past species, we come face-to-face with another problem. From evolutionary considerations, life probably began over 3 billion years ago as a single, simple species, probably resembling closely some present-day bacteria or blue-green algae. Since that time, an estimated 10 million or more species of plants and animals past and present have evolved (Simpson, 1952; Teichert, 1956). It is obvious that over the years one or more mechanisms must have resulted in an increase in the number of species.

These considerations underscore one of the basic concerns of systematics: How can one define a species? How many species are there, and were there? How and when and where were the various species formed? This area of systematic investigation is called *speciation.*

Phylogeny

The idea that species represent lineages poses an important question: What are the geneological relationships among these lineages? We are all familiar with the idea of lineages pertaining to individual members of our own family. How do these concepts apply to species rather than individuals as they occur in nature? It is obvious that these diverse species and the lineages they represent didn't arise individually and spontaneously. They must represent some sort of branching of ancestral forms.

To attempt to decipher these questions of origin, we find ourselves in the opposite situation of investigators trying to determine the fate of entities that may or may not be destined to be species. With these investigators, a view of tens of thousands of years into the future might solve their problem. In the case of the phylogenist, if record-taking man had lived as some sort of extracurricular observer for the 3 or more billion years that life has been in existence (longer to find out how it evolved), we would have a year-by-year blow as to how the present diversity of life had come about. But none of us did. As a result, we must devise techniques that will give us the most probable answers as to what did happen.

The possibility of such an approach, that is, a family tree of life, was not apparent to the forefathers of evolutionary theory. Lamarck gave a very short and cryptic family tree of certain facets of evolution, but Darwin and Wallace completely ignored a pictorial diagram concerning the evolution of the diverse forms of organisms. This was first done by Ernst Haeckel in 1866, who presented a remarkably good family tree for the major taxa known at that time (Fig. 1).

With such a new and revolutionary idea and tool as the family tree, one might suppose that all systematists would have begun expressing their views in this fashion, but with few exceptions this did not occur. Practically all of the exceptions were among paleontologists, notable investigators including Handlirsch working with insects and H. F. Osborn working with vertebrates. In both groups the fossilized parts (chiefly wings in the insects and skeletons in the vertebrates) were readily associated or comparable with living forms. Both structures presented series of changes that appeared to be gradual specializations in the evolution of the respective structures. The Osborn school set the stage for all the more recent systematic treatments of vertebrates. The Handlirsch school ran into opposition that is in ferment to the present day. The reasons for this disparity between insect and vertebrate phylogenies are simple. By 1900 the vertebrates treated by Osborn were fairly well known on a world basis, and Osborn dealt with skeletal features which have proven to give reliable clues to the phylogeny of the various vertebrate lineages. In the insects only a relatively small percent of the world fauna was known by 1900; Handlirsch was able to study only the wings satisfactorily and it has

Fig. 1.1 The phylogeny of living beings as conceived by Haeckel (1866) and expressed in a formal treelike diagram.

p   a   c   q

Plantae

Cormophyta

Anthophyta
Angiospermae

*Gymno-
spermae*

Pteridophyta
Lepidophyta
Rhizocarpeae
Filices
Calamophyta

Bryophyta
*Phyllo-
brya*

*Thallo-
brya*

Fucoideae
Sargassaceae
Laminariaceae
*Chordaria-
ceae*
11.

Florideae
Sphaerococcaceae
Ceramiaceae
10.

14.

**Chara-
ceae**
12.

Jnophyta
Lichenes
Fungi
13.

Archephyta
Ulva
Conferva
Desmidium
Nostoc
Codiolum.
9.

Protista

Myxo-mycetes
Physarum
Stemonitis
Lycogala
Trichia
5.

Spongiae

**Petro-
spongiae**
Siphonida
Ocellarida
Lymnorida
Bothroconida
Turonida

**Auto-
spongiae**
*Calci-
spongiae*

*Silici-
spongiae*

*Cerato-
spongiae*

*Myxo-
spongiae*

Rhizopoda
Radiolaria
Actinophryida
Acyttaria
Polythalamia
7.

**Myxocys-
toda.**
*Nocti-
lucae.*
6.

8.

Flagellata
Peridinium
Euglena
Volvox
3.

Diatomeae
Areolatae
Vittatae
Striatae
4.

Protoplasta
Arcellae
Gregarinae
Autamoebae
Amoebae
2.

Moneres
Protogenes
Protamoeba
Vampyrella
Protomonas
Vibrio
1.

Animalia

Articulata

Arthropoda
Tracheata
Crustacea

Vermes
*Anne-
lida*

*Rota-
toria*

*Scole-
cida*

*Infu-
soria*

Vertebrata

Amniota
Aves

*Mam-
malia*

*Repti-
lia.*

Anamnia
*Amphi-
bia*

Pisces

Amphirhina

*Monor-
rhina*

*Lepto-
cardia*

17.

19.

**Echino-
dermata**
Holothuriae
Echinida
Crinoida
Asterida.
16.

Mollusca
*Otocar-
dia.*

Himatega

18.

**Coelente-
rata**
Nectacalephae
*Petraca-
lephae*
15.

n ... n

14. 13. 12. 10'. 11. 9.   e   4. 5. 3. 1. 2. 7. 6. 8.   g   15.   16. 17. 18. 19.

Flori-
deae
Chara-
ceae
10. 11.

Fuci-
deae
10. 11.

Arche-
phyta
9.

Jno-
phyta
12.

Cormo-
phyta

Moneres

Flagel-
lata
1.

Myxo-
mycetes

Diato-
meae
5.

3.

Rhizo-
poda
7.

Proto-
plasta
2.

Myxo-
cystoda
6.

Spon-
giae
8.

Echino-
dermata
16.

Articulata
17.

Mollusca
18.

Vertebra-
ta
19.

Coelente-
rata
15.

13.

14.

Archephylum vegetabile   f   Archephylum protisticum   h   Archephylum animale

x ... y

Protista

Plantae   **Animalia**

I, Feld : p m n q *(19 Stämme)*
II, Feld : p x y q *(3 Stämme)*
III, Feld : p s t q *(1 Stamm)*
*stellen 3 mögliche Fälle der
universalen Genealogie dar.*

Radix communis Organismorum

Moneres autogonum

Monophyletischer Stammbaum der Organismen
entworfen und gezeichnet von
Ernst Haeckel. Jena, 1866.

s   b   d   t

since been shown that in these structures there is an astonishing amount of parallel evolution between different and often distantly related lineages. In my own opinion, Handlirsch did a remarkable job with the information available to him at the time he worked.

Probably because the early post-Haeckel phylogenies were based on groups having a fossil record, many containing unusual types no longer in existence, the idea became prevalent that phylogenies could be based only on fossil evidence. In the last few decades, however, techniques have been developed for working out highly probable phylogenies on the basis of living species only. New techniques include: the use of numerical methods, as used by Throckmorton (1968*b*) in the fly genus *Drosophila* and by Solbrig (1970) in the composite genus *Gutierrezia*; cytochrome C gene mutations, used by Fitch and Margoliash (1966); and polyploidy, used by Lewis in the genus *Clarkia* of the evening primrose family. If available, it is still extremely important to incorporate fossil evidence into the phylogeny, because fossils give us our best clues for tying a phylogeny into the geologic time scale.

As more and more phylogenies have been worked out, it has been discovered that the family tree provides a means of inferring past evolutionary events concerning the taxa under study. Examples of the events include the evolution of functional morphology, past dispersals of lineages from one part of the world to another, pathways of ecological evolution, the evolution of behavior and biochemical evolution.

Because of its potential in providing a better understanding of the past history of lineages, the study of phylogeny is another major concern of systematics. Like that of speciation, it is a highly theoretical field of investigation. It draws especially on comparative anatomy, evolutionary theory, and geology for its underlying matrix of theory. It seeks to answer especially the questions: What are the blood relationships between lineages past and present? When did their attributes evolve? What was their distribution in time and area? What was their ecological evolution?

Classification

In order for scientists to communicate with one another about species and phylogeny, they need to have names for the species and the various categories into which the species are grouped. Designating these names and arranging them into a system constitutes the area of effort called *classification*. Because of its usefulness as a storage-and-retrieval mechanism for biological information, classification is the third chief concern of systematics.

For biological classification as a whole we have devised a system of names and hierarchal categories of the nesting box type comprising what is called the *binomial system*. In this system each species has a name composed of two words. For example *Homo sapiens* is the name for man, in which the second word *sapiens* indicates man specifically and the first word *Homo* indicates the first nesting box, or genus, of the hierarchal system. The species name is applied to a scientific concept. The generic name and all others above it in the hierarchy of categories are applied to subjective groupings designed to provide what we think is the best system of boxes, slots, or pigeonholes in which to store our information about species or groups of species. The backbone of this hierarchal system of grouping species is the ascending order of categories beginning with *species* grouped into *genera,* genera into *families,* families into *orders,* orders into *classes*, classes into *phyla*, and phyla into *kingdoms*. Each member genus, family, order, etc. of one of these categories is termed a *taxon* (pl., *taxa*).

There are probably two million living species so far known plus tens of thousands of fossil species. To devise a classification that will provide names for all these and the many groups to which they are assigned is an extremely difficult task. To ensure as nearly as possible that the names are unambiguously understood by scientists throughout the world, complex sets of rules have been devised to regulate the formation and publication of names. To make the classification workable, identification aids of many types have been published, including descriptions, illustrations, keys, catalogs, and other publications that will enable the systematist to identify specimens at hand and determine to which part of the classification they have been assigned.

THE SPECIAL CONCERNS OF SYSTEMATICS

Based on the foregoing discussion, we can define systematics as the field of biology having the following special concerns about species:

I. *Biosystematics*, aimed at achieving a better scientific understanding of species and the lineages they represent, including:

  A. *Speciation*, establishing in an evolutionary sense the number of species representing distinct lineages in any transect of geologic time, the factors responsible for the origin of these species, and character change through time in any one lineage; and

  B. *Phylogeny*, establishing the relationships of these lineages and following the peregrinations of these lineages through time and space, biologically, geographically, and ecologically.

II. *Classification*, the formulation and improvement of an information storage-and retrieval system, including the ordering of species in a hierarchal classification of taxa plus the necessary devices for its operation. The totality of these classifications and their identification aids is the filing system used by all scientists for storing and retrieving information about all living things by taxa.

In the above listing of areas of study, the word *taxonomy* is not utilized. This term is widely used by both botanists and zoologists as a synonym of systematics. Taxonomy is regarded by some (Simpson, 1961; Mason, 1950) as essentially what I am treating as classification (see Chapter 11). However, Heslop-Harrison (1953) and others consider taxonomy to be the more inclusive term and systematics to be a subdivision of it (see review in Ornduff, 1969). In view of these discrepant opinions, I am considering that (1) taxonomy is a synonym of systematics and that (2) systematics has the majority vote as the inclusive name to be applied to the total field of investigation we are discussing.

The above listing also gives an expanded definition for the term *biosystematics*. It was originally proposed as the term *biosystematy* by Camp and Gilly (1943), defined as an attempt to "delimit the natural biotic units and to apply to these units a system of nomenclature adequate to the task of conveying precise information regarding their defined limits, relationships, variability, and dynamic structure." They felt that the binomial system of nomenclature did not express the evolutionary aspects of speciation then being realized. Their solution was to supplement the term species with an enlarged set of categories that would express the evolutionary origin of the classified entities. For example, they proposed the term *agameon* or *agamic species* for species that reproduced apomictically, *alloploidion or alloploidic species* for those arising by allopolyploidy, and *homogeneon* or *homogenic species* for those that are genetically and morphologically homogeneous, all members being interfertile. The suggested procedures were not generally adopted. Instead, the derivative term *biosystematics* came into use chiefly in botany for systematic studies emphasizing experiments with living material and genetics and cytogenetics (Solbrig, 1970). I am using biosystematics here as the investigational field of systematics based on any scientific information that can be brought to bear on the problems of the evolution of species, whether they concern speciation or phylogeny.

2

Investigating
Systematic Problems

The systematist strives to define and circumscribe the various species in his area
of study and to determine their phylogenetic relationships. The method employed is basically the study of comparative attributes interpreted in the light
of evolutionary and genetic theory, and, where relevant, in the light of geological and ecological theory also. Using this amalgam of scientific disciplines,
it is often possible to explain the origin and evolution of the various lineages
that are delinated.

In the last few decades, information from gross morphology, behavior,
and cytogenetics has been augmented by modern tools and new techniques
of observation. These include the electron microscopes, electrophoresis,
gas chromatography, ultraviolet spectrography, and high-speed computers,
to name only a few.

The ultimate goal of systematics is improving ideas and concepts concerning species and their phylogenetic relationships, or testing our previous
conclusions about them. Improvement may take many forms. We might
find that different populations each previously considered to be separate
species are only one, or that different populations previously considered to
be a single species actually represent many species. Exploration for new
characters might indicate that our ideas of the phylogeny of a group were
faulty and that a different blood relationship existed. These same new
characters might allow us to improve our classification of the organisms
studied, resulting in more accurate identification aids and a more useful
arrangement of the taxa.

BACKGROUND INFORMATION

Such words as "test" and "improve" imply the existence of background information expressing previous conclusions about systematic data. This
implication is correct. We already have devised classifications for all known

13

kinds of organisms and have detailed species studies and published phylogenies for a number of taxa.

This background or starting point of systematic information is actually tremendous, consisting of a voluminous literature and hundreds of millions of collected specimens deposited in many systematic collections. A recent survey (Anonymous, 1971) shows that in the United States the 20 largest systematic collections contain over 126 million specimens. Other institutional and private collections together certainly contain as many. The numerous large collections in Canada, Australia, Europe and other countries would undoubtedly bring the total world museum collections to well over a billion. The tremendous literature and number of specimens available for study are the heritage of the individual systematists who in the past 2500 years have accumulated material and recorded their observations and opinions on the identity, geneology, and classification of the species they studied. The job of this and following generations of systematists is to expand every field of these investigations into a more thorough and interpretive body of knowledge.

Surveying the huge volume of published books and papers in systematics, one might think that either little remained to be done or that routine tidying up was the sole task left, but nothing could be further from the mark. Although over a million and a half living species are known, it is likely that at least another million or two remain to discovered. Simply describing these additions and fitting them into the existing classification will be an immense undertaking.

The great opportunity for improving systematic knowledge and concepts lies in the realm of biosystematics. In a very few groups, notably the four-footed vertebrates and certain genera of higher plants and insects, studies of speciation have progressed to a remarkable point, but in most other groups our knowledge in this area is fragmentary. Studies of phylogeny are far behind those of speciation, in spite of the fact that phylogeny holds the key to the interpretation of practically all comparative biology.

The reason for this lag in the study of biosystematics is simple. It is relatively easy to collect and sort specimens into an arbitrary set of groupings and arrange these into a classification, but when we probe into matters concerning biosystematics we are asking questions that can be answered only by some knowledge of the past. This is because (1) the species we discover in our studies arose from previous ancestral forms, (2) the functional, morphological, and physiological relationships involving characteristics of species evolved from other precedent conditions, and (3) the patterns of geographic and ecological distribution we discover arose from older patterns.

Getting answers to these questions would pose no problem if we had a completely documented and easily understood record of all previous life. But we don't. Instead we must try to reconstruct what happened.

Biosystematics thus becomes an exciting exploration into the secrets of the past, the explanation of the present, and to some extent the prediction of the future of the biological world, charged with all the suspense of the Sherlock Holmes mystery story. Indeed, the ingredients of the biosystematics and "Holmes" detection methods are remarkably alike. If comparable investigational steps in the two methods are tabulated, the similarity is striking:

Examples of Operational Criteria

Investigational steps	Holmes method	Biosystematics
Recognizing the existence of a problem	Irregularity such as someone stabbed, jewels missing	Discrepancy in character combinations, ecology, inter-breeding, or geographic distribution
Gathering evidence	Questioning witnesses and suspects, securing fingerprints	Collecting specimens, making genetic tests, examining characters
Making inferences	Establishing guilt	Determining probable answer

RECOGNIZING A PROBLEM

Before a problem can be solved, it must be recognized. A problem is usually recognized when some observations or conclusions do not jibe with expected patterns. In a Sherlock Holmes episode it may be that a safe expected to be locked and full of jewels is found open and empty. In systematics it may be that two individuals thought to be the same species are noticed to exhibit striking differences in certain previously unobserved characteristics.

This sounds as if recognizing a problem was simple and straightforward. Actually it has several independent but nonetheless interdependent mental and observational facets. First, the investigator must be interested in some area of study aimed at classifying or ordering a certain part of our knowledge about living things. The initial interest may have originated in diverse ways. He or she may have had an early personal interest in plants, insects, or other organisms, and may have collected butterflies, snakes, shrubs, or mushrooms in some particular area. Next may have come a curiosity as to the names of the different kinds recognized in the collections, leading to efforts to find identification aids and other information about the group under study. The original interest may have been quite apart from studying a particular group of organisms from a strictly systematic sense. The investigator may have been observing the fungi attacking certain hosts, parasites attacking certain insects, the egg-laying habits of fish, the sounds made by crickets, and an endless list of other items.

Second, the investigator must learn the expected patterns into which observed data are thought to fall. This learning normally combines personal observation with the findings and conclusions of others which may be gleaned from discussion and reading.

Third, the investigator must be ready to notice when things do not "jibe." Careful study of the collected material may indicate that what appear to be two or more distinctive kinds all key out in the literature to only one species. The immediate question is: are the segregated entities only variants of one species, or do they represent one or more hitherto un-recoginzed species?

A person investigating insect distribution may notice that, contrary to expectation, a spring-appearing species A matured earlier to the north than to the south (in the Northern Hemisphere). This would lead to the question: Are the northern and southern populations the same or different species? An investigator of pathogens may note that a certain type of fungus produced larger spores on one host than on another, The obvious question: Is the difference due to the physiological effect of the host or does it indicate that two species are involved? An investigator of plants may note that species A, at that time considered a phylogenetic sister group of species B, had certain peculiar characters in common with species C. The obvious question: Is species A really more closely related phylogenetically to species C rather than to species B? An entomologist making new associations of immature and adult stages may note that on the basis of adult characters species A and B form one group, C and D another, whereas on the basis of larval characters species A and D fall together and C and B form the other group. Here the question is: How do we reconcile these different groupings based on evidence from different life history stages?

Whatever the question, once it is formulated we are able to take the next step in trying to find an answer. That step is gathering all available evidence.

3
The Material Basis of Systematics

We now come to the second investigational step, gathering evidence. In biological systematics anything we use as evidence must be what we consider facts. Most of our facts are the result of visual, auditory, olfactory, or tactile stimuli acting on our sensory organs and producing a mental image in our brain. Essentially, facts represent something in the external world that, at a particular time and under particular conditions, we can see, feel, smell, taste, or hear. Certain admissible evidence can be indirect, in that some modifying process enters between the presumed occurrence of something in the universe and our ability to sense that occurrence. For example, the transmitting electron microscope may record a visual pattern of a carbon caste of a specimen. In this instance the mechanics of the transfer of characters from the original specimen to the carbon caste can be duplicated with sufficient successive similarity that the character transfer process appears to be factual rather than haphazard. In the same fashion, sounds may be converted into visual images by the use of various electric or electronic devices. The same is true of the conversion of chemical concentrations, physical movements, and other characteristics of organisms into various types of visual signals such as kymograph strips, punched papers, or oscillograph records.

These bits of evidence, each one called a *datum* (pl., *data*), constitute the *material* basis of systematics. Opinions as to what they mean or might mean constitute the *hypothetical* or *theoretical* basis of systematics which will be discussed after the scope of the material basis has been elucidated.

CATEGORIES OF EVIDENCE

There are three categories of information into which evidence of possible use in solving systematic problems can be placed: (1) the characteristics of an observed individual, (2) its geographical location, (3) the ecological circumstances under which it was found.

Characteristics

These can be ascertained by observing individuals and recording the observations in a variety of ways, and conducting experiments or making observations in which two or more individuals are involved and recording various observed actions or interactions.

Observing Individuals

The types of observational evidence can be outlined conveniently around the headings *specimens* and *characters*.

Specimens. The first category of evidence comprises specimens or parts of specimens. In biological systematics, ideally one could wish for whole specimens but often this is impractical. A good-sized sequoia or banyan tree or a large whale, for example, would defy the full space facilities of many institutions. In such instances, the judicious preservation of selected parts of the specimen is the only practical solution.

It is impossible to preserve all the ontogenetic stages of one individual, because if any one stage is completely preserved, no further stages can develop. This ontogenetic impasse is resolved practically in several ways:

1.   by photographing or otherwise recording earlier stages and preserving the last one;

2.   preserving parts of the same individual at different stages of its life history (especially important with plants);

3.   preserving shed parts of stages of the same individual, such as the cast skins of the immature stages of arthropods; or

4.   preserving local population samples consisting of a random sample of one or more species or hybrids from a given locality. The sequentially preserved parts or the records assembled in the first three procedures indubitably represent one individual and one species. The fourth procedure may result in collections representing more than one species and must be considered questionable until supported by additional evidence.

We come now to the problem of the extent of collecting specimens. Considering the great diversity of species known to exist and the much greater diversity which is suspected, as well as the demonstrated variability known to occur within some species, it would be most informative to collect or at least examine every individual in the world. But limitations on time, space, and resources allow us to accumulate for study only an infinitesimal fraction of this number. If the species criteria are conspicuous to the eye and known to the collector (as in many plants, vertebrates, and larger invertebrates), 25 specimens of a local population sample of one species usually give adequate information. If two species and their hybrids are present, 50 to 100 specimens may be required. If the species criteria are not readily visible to the collector, as large a local collection as possible is indicated. To cite an example, Wheeler (1942) collected over 29,000 specimens of small leafhoppers of the genus *Empoasca* (whose identity can be established only by examination of macerated preparations of the abdomen) at light traps on the Arlington Experiment Farm near Washington, D. C. Laboratory studies disclosed that the collections contained 30 species, six represented by only 1 specimen each, five by only two, and six others by less than 20 (Table 1). It is possible, but not probable, that one additional collected specimen would have added another species to the list. Probabilities, however, deal with what we might expect rather than definitive actualities. Only additional data, often in large quantities, will give us insight into the ultimate scheme of things. Such a quest approaches closely, at least in a taxonomic sense, an examination of the totality of the living world. Thus the only guidelines we can formulate concerning extent of collecting are both broad and vague; we should endeavor to obtain and examine as much material as possible from as much of the world as we can reach. In this way we can hope eventually to identify all the species of living things and determine their total distribution. In practice, each systematist has special interests in a certain group of organisms or in the organisms of a special area. These investigators concentrate their efforts in their fields of interest, hence our systematic accumulations for the world biota as a whole are spotty.

fabae	24953	*atrolabes*	66	*curvata*	2
erigeron	2685	*obtusa*	36	*incida*	2
pergandei	558	*patula*	20	*pallida*	2
alboneura	239	*maligna*	13	*radiata*	2
solana	201	*chelata*	12	*ellisae*	1
recurvata	199	*dentata*	12	*luda*	1
sativae	194	*ditata*	9	*smaragdula*	1
bifurcata	140	*unica*	8	*torqua*	1
birdi	82	*delongi*	7	*trifasciata*	1
salicis	67	*copula*	2	*vergena*	1

Table 1. Numbers of identified leafhoppers of the genus *Empoasca* taken in two light traps on the Arlington Experiment Farm, from April to December, 1932-1941 (total, 29,517). [From Wheeler, 1942.]

Museums and herbaria in many parts of the world contain a voluminous quantity of preserved material representing practically every group known to man. Dealing in conservative figures, these museums contain over a billion preserved specimens of insects, other invertebrates, vertebrates, lower plants, and higher plants (Cohen and Cressey, 1969). It is difficult to preserve dead material of many microorganisms, and for these several living type culture collections have been established and are being maintained in 12 or more countries. Here one may have access to many thousand cultures of bacteria, fungi, viruses, algae, and protozoa (Clark, 1969).

Collections of living plants and animals in many parts of the world give the scientists access to living specimens of many thousands of species. Because of greater ease of maintenance, the plant collections are by far the more extensive. The most famous is that of the Kew Botanical Gardens, near London, England. For animals there is no counterpart to Kew, but several zoos (Whipsnade in England; Brooklyn, Brookhaven, and San Diego, in the United States) contain living specimens of the larger animals.

Characters. When we look at a specimen, even a small one, it is impossible for us to absorb all of its details simultaneously. For this reason we normally talk about a specimen in terms of the specific details or characteristics that emerge as the result of a "single look," such as its length, shape, the number of petals in its flower, or the number of hairs on its leg.

There is often ambiguity concerning the definition of the terms *character* and *characteristic.* Here I am considering a *character* to be some defined attribute of an organism, such as a particular bristle on a fly, the leaf of a plant, the shape of a tree, the snout of a mammal, and so on. A certain character may not be the same in different individuals, for example, the leaves in one individual may have a smooth margin, whereas those of another individual may be notched or serrate. In this case the character *leaf* has two (or more) conditions or states. The postorbital bristles of a fly may be present or absent, thick or thin, long or short, and so on. These various alternate observed conditions are the various *states* of the character *postorbital bristle.*

A particular character state occurring exclusively in certain specimens or species is *characteristic* of (or a characteristic of) a certain specimen or sets of specimens. Thus concerning the character *tail, bushy tail* is a characteristic of squirrels and *scaly tail* is a characteristic of rats. *Bushy* and *scaly* are different states of the character *tail.*

Delineating character states involves a comparison of the data obtained from observations of many specimens. The first step in obtaining this type of evidence is therefore to catalog and describe characters of individuals, then to correlate this information in a variety of ways to learn the different states of each character and the distribution of these character states in the population. This information gives an insight into the *genetic variability* or the *heritable characteristics* of a population or species.

The characteristics of an individual comprise all the attributes of the organism. They are sometimes defined as the expression of the entire genetic complement. This definition, however, is incomplete because the genetic complement itself is definitely an attribute or property of the organism. The total complement of characters of an organism is therefore best defined as the totality of its genetic complement and the expression of all its effects.

The genetic complement. This comprises the genome (the DNA in the nucleus) and the plasmon (the DNA found in mitochondria and other cytoplasmic elements of the cell). It is a presumed ideal that if we knew the DNA composition of all species, the course of evolution would be apparent. Even if we had this information, it is likely that problems of interpretation would arise. Nonetheless, many investigators have tried to find methods of determining the nucleotide sequences in DNA. Sonneborn (1965) has pointed out that this can be done by transcription from transfer RNA, which he did for the alanine transfer RNA determined by Holley *et al.* (1965). He pointed out that this was the first complete specification of the nucleotide sequence of a gene (Fig. 1).

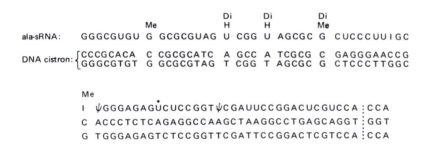

Fig. 3.1 First fully specified cistron or gene, for alanine transfer RNA. The RNA is shown in the top line, the DNA cistron below. Symbols in ala-sRNA represent the 3'-phosphates of: G, guanosine; C, cytidine; U, uridine; MeG, 1-methylguanosine; A, adenosine; DiHU, 5,6-dihydrouridine; DiMeG, N^2-dimethylguanosine; I, inosine; MeI, 1-methyl-inosin; 4, pseudouridine; T, ribothymidine. U* is a mixture of U and DiHU. The A,C,G, and T in the DNA cistron have the standard meanings of deoxyadenylate, deoxycytidylate, deoxyguanolate, and deoxythymidylate, respectively. (From R. M. Sonneborn, "Nucleotide sequence of a gene: first complete specification," *Science*, 148: 1410. Copyright © 1965 by the American Association for the Advancement of Science and reprinted with their permission and that of the author.

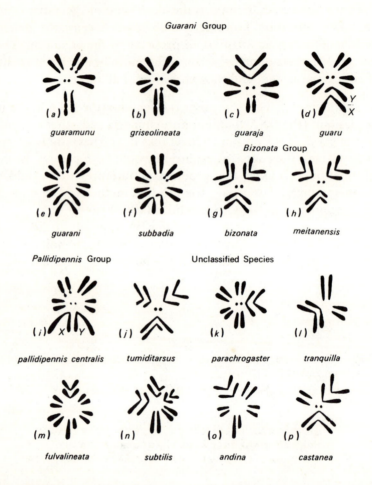

Fig. 3.2 Differences in number and shape of chromosomes in various species groups of *Drosophila*. (From J. T. Patterson and W. S. Stone, *Evolution in the genus Drosophila*, The Macmillan Company, 1952. (Copyright © 1952 by the Macmillan Company and reprinted with permission.)

It might appear that we could work backwards from the proteins to the RNA-DNA nucleotide sequences, but at the present time this proves to be impractical. As many as six RNA codons will bind the same amino acid in a polypeptide sequence; four RNA codons is the usual number; and only two code for a single amino acid. In a protein having as few as 60 amino acid residues, the number of possible combinations of DNA nucleotide arrangements responsible for the coding of it would be at least 4^{60}, almost arithmetic infinity. In view of this, other practical methods of assessing the DNA complement must be used.

Two methods are currently in use for studying DNA chemically (DeLey, 1969). First is calculating the molecular weight of DNA per cell. The molecular weight of DNA in selected species of bacteria varied from 0.2×10^9 to 3.1×10^9. Values calculated by various investigators varied within a range of about 60%, indicating differences in either techniques or strains.

Another measure of difference between the DNA complement of different species can be expressed as the molar percent of two of the nucleotides, guanine and cytosine, or the GC percent, to the total molar weight of all the nucleotides. This percent varies in bacteria (Lee et al., 1956; Belozersky and Spirin, 1960; DeLey, 1969) from less than 30% to over 70%. The known values in certain multicellular groups, each of which embraces a remarkable morphological variety, include: mammals, 39-45%; birds, 41-44%; fishes, 41-44%; insects, 41-43%; dicotyledons, 38-48%; and gymnosperms, 34-41%. When these figures are compared with those of 35-63% in the yeasts and 43-74% in the ascus-forming fungi, it is apparent that correspondence in GC percent values is not indicative of complexity of evolutionary development.

The DNA complement at present can be studied visually only in the nucleus, where it forms the chromosomes. By using various techniques of dissection and staining, the number, shape, and banding of chromosomes can be determined. In various taxa of both plants and animals, remarkable differences may be found on all three counts (Fig. 2).

Products of the DNA. These comprise the entire phenotype, and include a great array of chemical compounds, structures, and certain poorly-understood intangibles such as behavioral characteristics. Technical advances of the last two decades have greatly increased our ability to detect and record these products with greater detail and precision.

Chemical products vary from simple compounds such as salt crystals in the cell to proteins of great complexity — the cytochromes and haemoglobins, for example. Various types of paper chromatography, electrophoresis, and gas chromatography allow the detection and identification of extremely minute samples of many compounds that are proving to be useful systematic

Fig. 3.3 Scanning electron microscope picture of the diatom *Trinacria regina*, 565 X, showing the three-dimensional possibilities of the microscope. (Illustration kindly furnished from the Turtox Collection.)

tools (Alston and Turner, 1963*a*). Currently the flavonoids, terpenes, sesquiterpene lactones, and seed proteins are of special interest in botany, the haemoglobins, pheromones, venoms, and alarm substances in zoology.

Structures may be observed with regard to shape, color, texture, and other attributes. The naked eye can express many of these satisfactorily. If greater magnification is desired, various powers of light microscopes (from hand lens to phase contrast) give enlargement of from two or three to about 2000 diameters greater than the naked eye. Electronic magnification (chiefly the two types of electron microscopes) increases practical magnification to values up to at least 250,000. The scanning electron microscope combines remarkable depth of field and three-dimensional micrographs with high magnification and ease of preparation (Fig. 3).

Behavioral characteristics, such as rate of wing beat, physical features of songs, and mating movements may be recorded by various mechanical and visual methods including kymograph tracings, sonic recordings and their various graphic representations, and movie camera sequences. Modern instrumentation allows comparative studies on rates of photosynthesis, respiration, and transpiration.

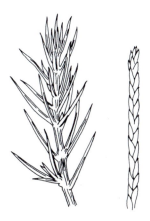

Fig. 3.4 Young (left) and mature (right) foliage of *Juniperus virginiana* at a magnification of 5 X.

Ephemeral characteristics pose a special problem. These include attributes that cannot be preserved such as many colors and tints of flowers, many insect and fish colors, texture of leaves or floral parts. Data concerning these phenomena should be recorded with the collecting data. The same applies also to characteristics that are impractical to collect adequately such as the bark of trees.

Ontogenetic changes. In all individuals, characters and patterns of characters change during development. In the simpler unicellular organisms these changes may be chiefly those of chemical composition and body proportions. More obvious changes occur in such groups as the higher plants in which the individual goes through quite different morphological stages — seed, seedling, mature foliage, flowering, and fruiting. In each stage many new characters appear and many or most of the previous characters may be lost. In plants, the leaves may change ontogenetically. In the genus *Juniperus,* for example, the leaves progress from the seed leaves or cotyledons to projecting needles to the plaited scale-like leaves of the mature tree (Fig. 4). In certain parasitic animal phyla such as the Acanthocephala or thorn-headed worms, the individual

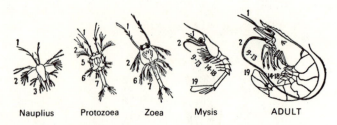

Nauplius   Protozoea   Zoea   Mysis   ADULT

Fig. 3.5 Metamorphic life history stages of the shrimp *Penacus*, showing the succession of changes in body form and appendages (1 - 19). (From *General Zoology*, 4th ed., by Storer. Copyright © 1951 by McGraw-Hill Book Company and used with their permission.)

passes through one or more juvenile stages before the adult stage is reached. The same is true of mites and ticks. Even more conspicuous examples of on-togenetic change are found in organisms having unusual polymorphic stages. Excellent examples are provided by the shrimps (Fig. 5), whose development embraces several distinctive juvenile stages, each separated by a molt. The most extreme types of ontogenetic polymorphism occur in the higher insects. Well-known examples include the butterflies, having egg, larval, pupal, and adult stages. In the last three stages, legs, antennae, eyes, internal organs, and other structures are remarkably different in each stage.

In these and more complex ontogenies, any one stage exhibits not all the characters of the individual, but only those characters occurring at one par-ticular point in its ontogeny. The characters of the individual embrace the sum total of *all* characters of *all* stages. Hennig has termed this sum the *holomorphology*. Each character of each stage of this grand total is a datum of evidence for problem solving in systematics. There is neither philosophical nor logical reason to make *a priori* judgments that any one of these characters is unimportant from the standpoint of giving us a basis for drawing a scientific systematic conclusion.

Because there have been many assertions in the literature that only this stage or that was important in classification, some defense of the present stand is indicated. One of the best examples concerns the screw-worm fly *Cochilomyia hominivorax* and its relative *C. macellaria.* It had been held by specialists working only with adults that these two were the same species because the adults were inseparable; the field scientists with control responsi-bilities maintained that two species were involved which were separated by morphological and behavioral characteristics of the larvae. After a twenty-year feud, the field observers were proven right, and eventually restudy of adult material unearthed differentiating characters for this stage also (Hall, 1948).

Another problem in assessing the holomorphology of a species is encountered when the organism passes through multiple generations: the characters of a *life cycle* vs. those of an *individual.* In the above discussion of individuals, we were dealing with a situation in which polymorphism arose in the development of one propagule into one adult. Many species of both plants and animals have polymorphism of another type, involving alternation of generations. A simple example is found in certain aphids or plant lice. The overwintering egg develops into a wingless generation *A,* which reproduces and gives birth parthenogenetically to embryos that develop into generation *B.* This generation in turn produces a succession of parthenogenetic generations, some having wings *(C),* some not *(D);* the late fall generations give rise to winged males *(E)* and wingless females *(F)* which mate and produce the overwintering eggs. Each form from *A* to *F* is slightly or markedly different morphologically, behaviorally, and presumably biochemically.

Remarkably interesting examples are afforded by plants. Certain algae and fungi have alternations of generations that are strikingly dissimilar, yet are obligate in the completion of the life cycle of the species (Fig. 6). In the wheat-rust fungus and its relatives, the life cycle consists of five different successive generations, each producing a distinctive type of spore; included in the cycle are also alternations of host plants.

In organisms in which there occurs an alternation of generations we have a compounding of the "holomorphology" of those organisms in which no such phenomenon occurs. From the standpoint of assembling evidence concerning phylogenetic problems, the concept of "total characters of the individual" must therefore be expanded to include the "total characters of the life history."

Dioecious species. The same problem applies equally to dioecious bisexual species. In these the two sexes are in different individuals (dioecious, translated literally, means "two homes"), including many plants (e.g., *Ghinkgo* and *Ilex)* and almost all multicellular animals. In the life cycle of these dioecious species from fertilized egg to fertilized egg, two contemporaneous individuals are necessary, a male and a female. Frequently the two sexes differ greatly in many characteristics (Fig. 7). In all such instances, the total characteristics of the life cycle are the sum of those of the male and female individuals of

...erience has shown that preconceived notions con-
...one sex vs. the other in making systematic conclu-

...n is found in animals such as ants and termites hav-
...ich many individuals of a colony do not reproduce.
...g reproducing and nonreproducing individuals is

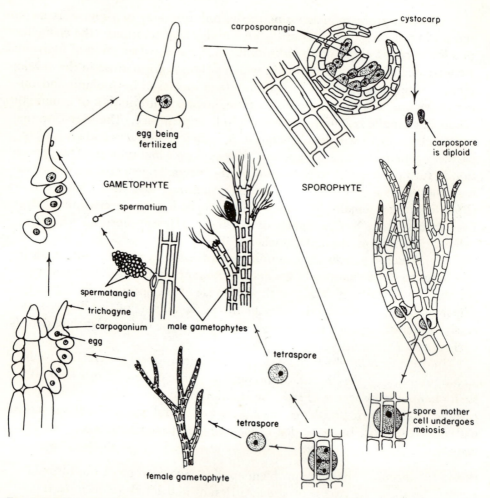

Fig. 3.6 Life cycle of the red alga *Polysiphonia*. (From H. T. Northen, *Introductory Plant Science*, 3rd ed. Copyright © 1968 by the Ronald Press Co. and reprinted with permission.)

equivalent to a bisexual nonsocial pair. Invariably each caste of the species has a distinctive morphology. The holomorphology of the life cycle is therefore the sum total of the characters of each caste and sex.

The totals and grand totals we have been discussing are the characters of the species; in short, the species genome and all its chemical and physical effects.

Ecophenotypes; ecological variation. Many instances are known in w̶ same genotype produces a different result under different circums̶

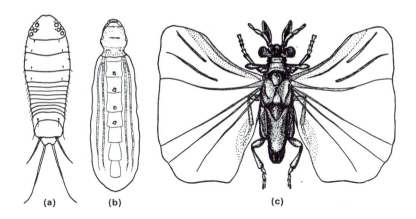

Fig. 3.7 Stages of the life history in the Stylopidae (Strepsiptera): (a) minute early larva of *Xenos pallidus,* enlarged over 50 times more than (b) and (c); (b) mature female of the same; (c) mature male of *Stylops pacifica.* (From Bohart, 1941.)

differences are nongenetic and are called *ecophenotypes.* They may result from either differences in ecological conditions or situations involving allometry.

Ecological conditions. Undernourished individuals are smaller than normal, and this smaller size may be accompanied by changes in other body proportions such as fatness or thinness. In plants, individuals in shady situations may have larger and thinner leaves than those in sunnier spots.

More drastic well-known differences occur in aquatic plants such as *Potamogeton* and *Ranunculus,* in which the leaves borne below the water surface are finely dissected, and those borne above the water are much less or not at all divided. Njoku (1956) found a comparable effect produced by sunlight on leaves of the morning glory *Ipomoea coerulea.* Leaves in direct sunlight were heavily lobed, those in shade had few or no lobes.

Conspicuous polymorphism occurs in certain Old World migratory locusts: individuals occurring in uncrowded conditions are robust, short-winged, and flight-less; those occurring under crowded conditions are slender, long-winged, and powerful fliers. The two phases differ in other morphological features. In crickets of the genus *Achaeta,* polymorphism in growth features and wing length is caused by temperature and dietary protein level (McFarlane, 1964; Mathad and McFarlane, 1967), and male albinism is produced by an overdose of vitamin E (McFarlane, 1972). Some dramatic ex amples involve color. In 1924, Knight demonstrated that in the stink bug *Perillus bioculatus* both temperature and humidity have a profound effect on body color. Low temperature and high humidity produce almost black body color, high temperatures and low humidity produce yellow body color; intermediate conditions produce orange body color. Biochemical evidence revealed that these differences result from the rate of metabolism of the organism. If low, the metabolite melanin accumulates in the body wall, giving a black color; if higher, the melanin is converted to carotin, giving an orange color; if still higher, the carotin is converted to flavin, giving a yellowish color. Furthermore, it was determined that low temperature and high humidities decrease metabolic rates, the opposite increase them. These differences are classical examples of ecophenotypes. These effects are not general throughout the insects, but are observable as climatic-associated differences in a large number of groups.

Formerly it was thought that all different growth forms of a single species of plant were due to the effect of the environment on the same genotype. Thus tall forms growing in moist, shady areas and short forms growing in drier open areas were considered to be simple ecophenotypes. Using cultivation experiments with the hawkweed *Hieracium,* Turesson (1922) demonstrated clearly that many of these forms were correlated with different genetic constituents, each adapted to a specific ecological situation. His work demonstrated for the first time the widespread occurrence of intraspecific, habitat-correlated genetic variation as opposed to the ecophenotypic expression of a similar genome.

Comparable differences between high elevation and low elevation plants had also long been laid to an ecological correlation — high elevation (low temperatures) populations are short and tundra-like, low elevation (higher temperature) populations are taller and more shrub-like. In his investigations of Californian populations of the yarrow *Achillaea borealis,* Clausen (1951) found that the tall lowland form differed from the dwarf alpine form in genetic composition, and that it was a difference in genotype that was responsible for the different growth forms occurring at different altitudes. Such different races, adapted genetically to different climatic situations, have been termed *ecotypes,* indicating that they are not simple automatic ecophenotypes but owe their differences to a genetic basis.

The above examples of ecotypes represent experiment rather than observation, but they do stress the necessity of sampling for character variation at frequent intervals along ecological gradients. If intergrading character states follow an ecological gradient, then various transplant tests are indicated to determine if the different variants are ecophenotypes or ecotypes. One of the best techniques is to raise cloned individuals in different environments. Each member of the clone is genetically identical; hence a series of sister clones in different environments will indicate any ecophenotypic expression.

Systematists have recognized that in many animal groups certain temperature-humidity effects are of general application and have expressed the relationships involved as various rules (Allee and Schmidt, 1951). Bergman's Rule states that body size increases with a decrease in climatic temperature. Allen's Rule states that various appendages and extremities of animals are relatively shorter in cooler climates. Gloger's Rule indicates that melanism increases with decrease in temperature. The basis of each rule might seem to be that the same genotype produces a different product under different ecological conditions. This is not necessarily the case. These three and a few other similar rules may apply to species or to populations of the same species. If the former, it is almost certain that climatically-associated differences are ecotypes. If the latter, they could be either ecophenotypes or ecotypes. Only carefully planned experiments will give clues as to the correct solution.

It is possible that the expression of all genetic components of an organism are influenced to some extent by environmental factors but most of them to such a minor extent that the effect merges imperceptibly with what is considered as normal genetic variation. In some instance, however, the environment has such a powerful effect on the expression of certain characters that the various conditions produced appear to be highly distinctive characteristics. Striking ecophenotypes are especially common in plants. The same species may have different growth characteristics in different climates, as for example the short, contorted growth form of *Pinus contorta* in coastal British Columbia and its tall, erect form (the lodgepole pine) in the arid mountain areas to the east. Ecophenotypes may be produced by critical ecological changes occurring at different seasons. For example, winter females of the caddisfly *Sortosa distincta* have only vestiges of wings, but females maturing in summer have normal wings extending beyond the body; no intermediates are known and presumably some critical temperature or light intensity controls the extent of wing development. In certain other animals striking differences in size and color are controlled by environmental differences in temperature and humidity (see Allee and Schmidt, 1951).

One of the most striking examples of the effects of environment on structure was the demonstration by Müller (1954, 1958) that photoperiod controls the morphology of the aedeagus in certain leafhoppers. In *Euscelis plebejus* and *lineolatus* (Fig. 8), if the nymphs develop during a short-day regime, the tip of the adult aedeagus is slender and simple; if the development is during a long-day regime, the aedeagus is large and ornamented with lateral

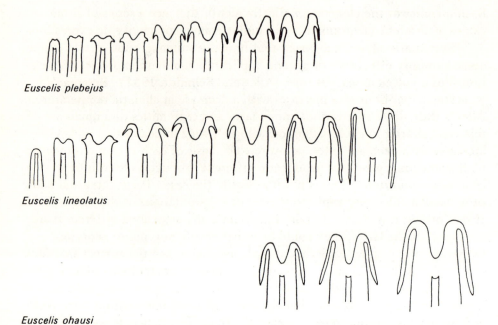

Euscelis plebejus

Euscelis lineolatus

Euscelis ohausi

Fig. 3.8 Seasonal variation of three leafhopper species of the genus *Euscelis,* caused by variation in day length. Shorter light period effects are to the left; longer ones, to the right. (From Wagner, modified by Müller.)

processes. These and intermediate phenomena produce a seasonal change of structure. Before this ecophenotypic photoperiod reaction was discovered, each species was considered to be four or five species on the basis of genitalic differences. In other species of *Euscelis,* different photoperiods seem to produce little difference in phenotypic structure.

A remarkably similar situation was discovered in the horsetail plants of the genus *Equisetum* by Hauke (1971). He found experimentally that in *E. arvense* higher light intensities of all wavelengths produce a disproportionately higher number of male gametophytes; that in *E. hyemale* red light produced the same higher male ratio in contrast with either blue, green, or white light; and that in *E. fluviatile* the sex ratios were unaffected by either light intensity or quality.

These two examples of differing photoperiod response in different species of the same genus, one example in animals and the other in plants, interject the thought that no generalization in any group should be made above

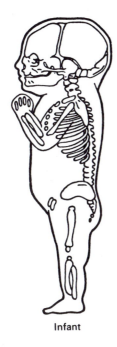

Infant   Adult

Fig. 3.9 Allometry in the human head and trunk, illustrating the differential rates of growth in head and trunk. (Modified from W. Etkin in *College Biology*, Crowell, 1950 as it appears in *Life, An Introduction to Biology*, 2nd. ed., by George Gaylord Simpson and William S. Beck, Harcourt Brace Jovanovich, Inc., 1965. Copyright © 1965 by Harcourt Brace Jovanovich and reprinted with permission.)

The species level. In other words, in any genus, each of the contained species may be a law unto itself as to its phenotypic expression with respect to light quality or quantity.

Other examples involving especially fungi and insects include morphs produced by the same species growing on different hosts or different parts of the same host. When different morphological types show such relationships, it is advisable to attempt transplant experiments of sibling individuals to test the possibility that they represent ecophenotypes.

Allometry. This results when different parts grow at different rates at different times in the ontogeny of the individual. In humans, for example, the head grows faster and the trunk slower in embryonic life, wheras after the birth the opposite is true. As a result, the head of a newborn human is much larger in proportion to the trunk than that of the adult (Fig. 9).

This allometric feature may produce marked differences in comparable stages of the life history. Examples are certain sawflies of the genus *Dolerus*. In small individuals, the head is narrowed behind the eyes, but in large individuals it is swollen and considerably wider than the anterior part of the head. In horses, Simpson (1944) demonstrated a similar allometric growth relationship between the length of the muzzle compared with that of the whole cranium. The muzzle length becomes relatively longer as the absolute size of the horse increases. In living species, this changing ratio progresses from foal to adult. Simpson plotted this muzzle/cranium ratio for the small, extinct horses of early Cenozoic to the successively larger horses that evolved later, and found that, with one small deviation, it agreed with that of the ontogenetic stage of living species. He concluded that in horses a genetic change resulting in different overall size would automatically produce a marked difference in the muzzle/cranium ratio.

Matsuda (1960) has since shown that many characters in certain aquatic insects thought to represent distinctive genetic determinants are probably only the allometric expressions of increases in body size.

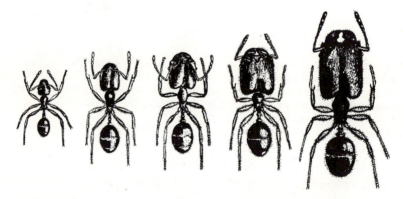

Fig. 3.10 Allometry in worker ants of the genus *Pheidole,* showing the increase in the relative size of the head with absolute size of the body. (From Wheeler.)

A phenomenon probably involving allometry occurs in certain insects having what are known as morphs. In many ants, the worker caste may range from small individuals having relatively small heads to larger individuals having relatively large heads. This phenomenon is especially marked in the ant genus *Pheidole* (Fig. 10). It was formerly thought that the large workers with large heads cracked the seeds brought to the nest by the more agile small workers. But more recent observations suggest a defensive role for the large-headed individuals. In a few species of thrips the same phenomenon occurs concerning the size of front legs. No function has yet been determined for this diversity of structure. With these thrips morphs there is still a question as to whether the different types are due solely to allometry or whether some type of genetic control is involved.

Behavior. From the standpoint of assessing the heritable characters of a species, behavior is a special problem. In both plants and animals, units of behavior may be the result of either heritable characters or ecophenotypic responses that are environmentally induced — in the latter category the concept of learning would be included. For this reason, we must exercise great care in thinking of all observed behavioral characters as heritable ones.

Genetic determinants of behavior have been discussed by Hirsch (1967) and Parsons (1967). The existence of large numbers of these determinants has been established by experimental methods, including such traits as social dominance in chickens, alcohol preference in mice, mating songs in frogs, and mating behavior in *Drosophila*. In plants, comparable examples are difficult to find but might include complex genetic mechanisms that inhibit self-pollination, permit certain mutants to live in certain types of soil, or that result in adaptations advantageous to living in varied environments such as cold-wet, hot-dry, or hot-wet, as in the millfoil *Achillea borealis* (Clausen, 1951).

Many examples of behavioral differences between species as yet unexplored experimentally also seem certain to be genetically controlled. For example, in certain species of pine sawflies *(Neodiprion)*, the female lays an egg in a pine needle, takes two steps forward, lays another, and so on. In other closely related species the female takes four steps between each egg deposition. In many insects (e.g., Kessel, 1955) and frogs (Jameson, 1955), a series of different patterns of mating have evolved in various families or genera, frequently species-specific, including the offering of bait, production of songs, and details of copulation. In termites of the genus *Apicotermes*, the blind workers construct an ovoid underground nest; nests of each species differ in intricate ways from those constructed by other species, and the different kinds form a convincing evolutionary series (Schmidt, 1955 *a,b*). In these examples there would seem to be little possibility of the individual learning its behavior pattern, especially in instances such as the termite example, in which the nest is the product of hundreds or thousands of workers.

There are, however, many observed examples in which differences in the behavior pattern of individuals are not due to genetic determination but to environmental causes or to learning. For example, in many crickets and cicadas the songs are so similar within the species and so different between the species that differences in the inheritance would seem to be reflected. Indeed, the native North American crickets of the genus *Acheta* were first recognized by their songs; to this day no other satisfactory diagnostic features have been found. But here again caution must be exercised. In many species of crickets, the tempo of the song is a function of temperature; hence songs of various species must be correlated with the temperature at which the recordings were made.

In the western and eastern meadowlarks, usually considered most distinct on mating-call characters, Lanham (1957) found that only the alarm "cluck" appeared to be a heritable behavioral difference between the two species. The mating and other calls were learned by young birds on the basis of association and mimicry, that is, learning. Thorpe (1961) discussed bird song phenomena further.

Working with blister beetles, Selander (1964) and Selander and Mathieu (1969) found that each species has a basic pattern of mating behavior, but that males having previous mating experience frequently omit some steps in this basic sequence and achieve a short-cut to copulatory activities. Here again, learning apparently produced a behavioral pattern different from the presumed basic genetic type.

In plants, certain behavioral characteristics such as date of flower setting are determined by environmental factors, including seasonal and circadian rhythms of temperature and light. Other behavioral characters, such as deciduousness, may be controlled by either genetic or environmental factors, or both.

Because of these uncertainties concerning the components of behavior, genetic versus learning or environment, the possibility must always be borne in mind that differences in behavior that have not been tested experimentally may not be expressions of the genetic code, but may be due to ecological or age factors, or to learning.

Homologous heritable differences. In the light of these circumstances, we must attempt to ensure two criteria in our use of character differences:

1. The differences observed should represent heritable differences and not differences due to the same genome developing under different environmental conditions; and

2. Compared differences should be those exhibited between comparable life history stages. One should not, for example, compare characters of a gametophyte fern individual with those of a sporophyte individual; those of a male bird with those of a female; or those of a larval insect with those of an adult. Compared characters between different individuals should be homologous both for structure or compound and for ontogenetic and life cycle stages.

Experiments with Individuals

Certain biological or biochemical characteristics of organisms are at present difficult to express as those of one individual, but may be expressed as differences between individuals. These differences can be tested by experiments involving two or more individuals, tests in which the *product* of the experiment is the measurable quantity.

Two experimental techniques involving reactions between individuals have proven useful in systematics — hybridization and serology.

Hybridization. In hybridization experiments, individuals of two sexes or types are crossed and the differences (if any) between the two parents are expressed as characteristics of the hybrid progeny (if any) resulting from the cross. Crossing experiments may involve whole individuals or only DNA or RNA strands.

Crossing experiments with whole individuals should tell us whether or not two selected individuals of opposite sex will or won't mate and produce offspring, how well any progeny will succeed, and the results obtained when such progeny are crossed with each other or with the parent types. Using the words "parent types" implies that we have two colonies established, either from different sources or of different recognizable types, and that we are drawing experimental animals from continuing laboratory populations.

Positive crossing experiments are indubitable facts, but negative results may not be. In the latter there is always the possibility that some condition of the experimental situation has introduced artificial barriers to successful reproduction that do not occur in nature. To ensure against this requires vigilance and ingenuity on the investigator's part. Failure or success in inducing mating in the laboratory are both questionable indicators of events in the field. Furthermore, even the production of viable laboratory hybrids gives no information about their ability to survive in the field.

Crossing experiments with DNA or RNA strands will give a measure of the nucleotide homology between various strands of different parentage. DeLey (1969) has outlined certain techniques useful in this operation and hints as to trouble-spots that may be encountered in its use.

Serological tests. These tests are aimed at getting a measure of the relative similarity of the total or partial protein composition of different individuals. To be useful, at least three individuals must be compared, such that it can be stated that individual A is more like individual B and less like individual C. With present techniques (Hunzicker, 1969), a remarkably precise set of interindividual comparisons can be established.

As with any experimental interindividual device, when conclusions are drawn from serological trials, it is necessary to understand what is observation and what is inference.

Geographic Records

The geographic distribution of living things contains a myriad of clues important to understanding past dispersal patterns of diverse lineages. Like behavior, geographic distribution does not express solely a heritable character of an organism It does express the survival value resulting from the heritable physiological characters of an organism interacting with the ecological conditions of the environment. The geographic distribution of any taxon is simply the qualitative factor of *where it occurs,* quite apart from the quantitative factor of where it may occur *abundantly* or *sparsely.* In this sense, the data of geographic distribution provide records of particular specimens occurring at exact localities that can be found and spotted on a map.

Several writers urge that these be expressed in units of longitude and latitude. This, however, is an ideal (if it is an ideal) that is not likely to become standard practice for some time, if ever. When we make collections, it is most convenient to use place names or localities we know or can find on an available map. It has almost invariably been so and I see no reason to expect much change in the future.

This raises the first major problem concerning the data of distribution. Place names may change or disappear, and older names in the records may be difficult or impossible to locate. To avoid this loss of good data, several avenues are open. First, we can build up collections of maps and learn to use the extensive map libraries being developed at many major library centers. Second, when material is being labeled or accessioned, care should be taken to orient the collecting site with a stable governmental unit such as a county, parish, or shire. Third, if material from a little known area is being recorded in the literature, a base map locating collection sites can be included in the report.

A second major problem concerns the total area covered by the records. What we hope to achieve by plotting our records of a taxon on a map is an accurate delineation of the total range of that taxon. When the records are plotted, we end up with two results:

1. A number of spots in which we know that collectors have worked with sufficient diligence to find specimens of the taxon, and

2. Unspotted areas (a) in which collectors have not worked assiduously enough to locate specimens if the taxon does occur there, or (b) in which the taxon does not now exist or did not exist at the time individual fossil sediments were deposited

There is probably no present or past species for which our plotted records will represent the absolutely complete range. Realizing that the only natural geographic unit is the entire world, we must therefore strive to obtain records from all pertinent parts of it. The word pertinent is inserted because of certain ecological considerations. It is obvious that oak trees will not grow out of an ocean, so for oaks we need collect only in terrestrial areas. But this type of reasoning cannot be carried too far or we are in danger of missing important data because we decided not to look in some unlikely spot. For example, it was assumed for years by stonefly collectors that this group of insects was poorly represented in the Ozark-Ouachita region, and little effort was made to collect there. When the region was intensively collected, its fauna was discovered to be unsuspectedly rich (Ross and Ricker, 1971). The more extensive and intensive the collecting, the greater is the chance that our maps represent something close to the actual range of a species.

To obtain a better basis for judgment concerning species ranges, some systematists record their negative records. This is not a perfect guide, but is much better than a completely unexplained void in a map. If these negative records represent careful searches under good collecting conditions, they increase the probability that voids in a map indicate voids in a range.

A third major problem concerning distribution is that a spot record is no better than the accuracy of identification of the material on which the record is based. If the basis of identification of a species is changed, all previous records may be either suspect or outright unusable pending a reexamination of the older specimens. If these are not available, the older records may need to be discarded *en masse* and a new start made. For example, in 1953 the leafhopper species *Empoasca solana* had been recorded over a wide range of localities in North and South America. In 1953, Young demonstrated that the material previously identified as *E. solana* was a composite of a dozen species. All earlier records of *E. solana* that could not be rechecked against the new set of diagnostic characters had to be dropped from scientific consideration.

Ecological Information

One reason for gathering data on ecological parameters is that, when it has been integrated with highly probable phylogenies, there may result the discovery of various facets of past evolution in relation to ecological factors. To obtain and record this ecological information is a formidable task. In the first place, when we collect our material we often don't know what ecological factors we might be interested in later. Secondly, many of these factors could not be determined at the time the collection was made. Suppose we were collecting mice throughout North America and because of expediency all collecting was done during the summer. Later we might want to correlate mouse phylogeny with the annual amount of snowfall. This latter would not be evident when we made our individual collections but would be available as a meteorological reconstruction.

A number of ecological parameters can be noted at the time of collection. For plants, information concerning soil, light or shade, moisture, and associated plants should be recorded. For animals, the ecological community in which specimens are collected should be noted. Of especial importance is the recording of the host for specimens that are parasitic, predaceous, or phytophagous, such as fungi, certain other plants, parasitic animal phyla, and many mites, crustaceans, and insects. Also, the position of parasites on or in the host is important. For internal parasites, the occupied organ or tissue should be noted; for external parasites such as lice or mites, the area of attachment should be recorded. Admittedly, the collector might not be able to identify the plant or other host, but could collect a sample of it and ask an expert to identify it. Conversely, a mycologist collecting fungi on insects could send host specimens to an entomologist for identification.

From the standpoint of ecological information, for most material the date of collection (month, day, and year) is an important item. Series of dated collections made throughout a year or season can be combined to determine seasonal development and adjustment; series of collections made during the same span of time or seasons but at different localities will indicate the seasonal adjustments of the species as influenced by different climates. The most important information arising from dates of collection may be associated with climatic factors such as unusual temperature or rainfall extremes. These we cannot record at the time of making the collection but the information can be obtained later through the use of climatological reports.

SUMMARY

These items concerning character states, geographic distribution, and ecological information comprise the material basis of systematics. These are the bits of available information that we can use for problem solving in systematics.

4
Scientific Reasoning in Systematics

At this point in our investigation we should have delineated the problem to be studied, gathered all available evidence pertaining to it, and be in a position to take the third step, the making of meaningful inferences concerning its solution or explanation. In actual practice, however, it is just at this point that we often encounter serious difficulties. These arise from a frequent lack of agreement concerning the solution. We find that:

1. In a few instances, the result does seem obvious, and all observers agree as to the solution of the problem.

2. In a few instances, no one is satisfied with any suggestions made by anyone else and many are reticent to propose a solution.

3. In most cases, certain investigators arrive at one answer, others propose one or more different answers. Which is correct? The result is usually a heated argument between proponents of conflicting solutions.

If we have considered all the available evidence, the only way to attempt a solution of such conflicting conclusions is to examine our thinking. Perhaps the differences of opinion are due to faulty thinking on the part of one or more of the disagreeing scientists. Several types of faulty thinking might be involved, including:

1. effect of preconceived ideas,
2. deficiencies in background knowledge,
3. errors in logic.

When we examine our thinking according to these lines, we invoke the principles of the scientific method.

The scientific method involves a particular kind of trained or learned way of thinking about evidence and its meaning. In many respects this kind of thinking is simply an extension of anyone's normal efforts to organize his or her experience. From infancy we try to organize the evidence we encounter. We associate *fire* with *hot* or *burn,* *hunger* with *lack of food,* *sun* with *warmth,* and *snow* with *cold.* We develop the idea that our sensed observations fall into an orderly arrangement by the previous examples as well as the following:

> *Chairs* are associated with *sitting.*
>
> *Leaves* are associated with *plants.*
>
> *Barking* is associated with *dogs.*
>
> *Purring* is associated with *cats.*
>
> *Singing* is associated with *birds.*

If the world were otherwise, if fishes and dogs occasionally sang like birds, if leaves sporadically grew out of lamp posts and teacups, we would be unable to made sense out of our observations. But most of the time we can. As a result, from our earliest days we build up the idea that there is a sensible correlation between the various sensed data that we hear, see, feel, taste, or smell. These data are our clues to understanding the universe. Interpreting this evidence should give us the possible answer about what the universe is, including atoms, energy waves, chemical compounds, species, and their origins.

The neurophysiological basis of our sensing and ordering is imperfectly known, but includes at least two parameters. First, stimuli from the external world excite sensory receptors and thus cause some sort of change in the physicochemical properties of the recipient. Second, the intellectual capabilities of the individual attempt to record and store inferential correlations associated with such physicochemical changes and produce some sort of deductive and predictive conclusions that will be helpful in future exposures of the individual to the same or similar external stimuli.

In the context of systematics, we note that cows give rise to calves, that grass seeds give rise to grasses, that cockroaches give rise to more cockroaches, and so on. We note also that a butterfly egg hatches into a larva; this grows and becomes a pupa, and the next generation butterfly emerges from the pupa. In both plants and animals there are countless other examples of an orderly progression of metamorphic types.

Because of this myriad of orderly concepts, even in the face of others that seem disorderly, we have an inherent belief that

1. the world is orderly, even though at times complex;

2. our senses will tell us salient features about it;

3. we will be able to ascertain orderly relationships about this world; and

4. we will be able to make logical inferences about this world.

In making inferences, relationships that seem orderly over a broad spectrum of observations are expressed as basic concepts, premises, or hypotheses, and these are used for deduction or prediction concerning comparable phenomena of the biological world.

INTERPRETING EVIDENCE

As pointed out by Randall and Buchler (1942), there are different ways of interpreting the sensed data that we call evidence. These different ways center around four common methods of thinking. The first is *intuitive*. In this method there is a belief that the human mind can arrive at the correct answer concerning any question about the universe by some inner mental activity without using evidence. According to this philosophy, the data should line up with the conclusions; if they don't, the assumption is that there is something wrong with the data. The intuitive method leads only to conflicting dogmatic statements with no possibility of using evidence to decide which might be right. From the standpoint of scientific problems, the purely intuitive method has never led to an advance in our knowledge.

The second method, suggested by Hull (1972), may be called *that of the heart*. This method is an emotional one based on the intensity of feeling a person may have concerning some outcome. He or she may say "This particular answer means so much to me that it *must* be true." In everyday life it may be a child asking if he is going to get a certain present for Christmas, or someone insisting that a word is spelled a certain way. In systematics, a person may say of a population possessing some unique difference, "This must be a different species" without looking at other evidence; or that "This genus must belong to a certain family" without making a critical evaluation of the phylogenetic inferences. In systematics, the emotional basis usually has its roots in the person's unwillingness to depart from the influence of earlier learning or from tradition.

The third common method of interpreting data is called *authoritarian*. In this method only part of the evidence is admitted as valid, and this is that part permitted by the authoritative body. This body might be the state, the family, the church, or various scientific bodies, cliques, or establishments. The idea is that the larger body has a collective wisdom greater than that of the individual and is therefore much more capable than the individual of knowing what evidence is informative and what is not. In short, the individual surrenders his right of independent thought to the majority. In science the authoritarian method has proved to be in error on occasions too numerous to detail here. Galileo and Scopes will serve as well-known examples.

The fourth common method of thinking about evidence is the *scientific method* to which all those who call themselves scientists are supposed to adhere. In this method all objectively sensed data are admitted as evidence concerning our queries as to the nature of the universe.

But admitting all evidence really opens Pandora's Box. It is not only possible but frequently true that various items of evidence are contradictory. Contradictory evidence leads to alternative answers, and we are faced with the necessity of trying to determine which answer is more likely to be the correct one. Our efforts may not seem to be conclusive, necessitating a search for even more evidence that might throw decisive light on the problem at hand. Even if all our evidence points to the same answer, we cannot forget the possibility that at any moment additional evidence may show our explanations to be wrong and necessitate a change or modification in our conclusions. Because of this, scientific explanations can never claim to be the absolute truth but must rather be expressed as a statement or hypothesis that appears to be true on the basis of the evidence before us. As a consequence, our hypotheses and theories for explaining phenomena can be expressed only in terms of the *probability* that they are correct. This probabilistic aspect of science was first postulated in the middle 17th century by the Dutch astronomer Huygens (Thiel, 1957).

Systematics is one of the few scientific fields in which the probabilistic nature of scientific results has frequently been misunderstood. Results obtained in phylogenetic and speciation investigations are continually being criticized because of the fact that new information or ideas can lead to changes in phylogenies or species concepts. And so the critics argue that such studies mean nothing because the investigators "can't be sure." The words "sure" and "proof" are incompatible with hypothetical scientific statements, but nonetheless we can often achieve a degree of probability remarkably close to proof with many of them. But we must *never* forget that additional evidence may either falsify or alter any such statement.

LOGIC AND REASONING

The practice of the group called scientists consists in a rigorous adherence to the type of formal logic and inductive-deductive reasoning that has been shown over the centuries to be most likely to avoid fallacious conclusions. It is not average or general or "to everyone's liking" — it may be to a few people's liking (Thouless, 1953; Latta and Macbeath, 1956).

Formal Logic

The logic is essentially the syllogistic logic of Aristotle, with its various rules to ensure that the thinking is kept straight. For example, one may observe

many robins nesting, always laying blue eggs. Upon seeing a robin starting a
nest, one can say with a high degree of probability that it will lay blue eggs.
But if one finds a nest having blue eggs, it is not logically safe to say that
these eggs were laid by a robin. This could be said only if one had ascertained
the color of eggs laid by all other birds inhabiting the area and found that
none of them ever laid blue eggs. In point of fact, catbirds, cuckoos, and
many other birds commonly occurring with robins do lay blue eggs. This
case is only one simple example of the many permissible and nonpermissible
permutations allowed by formal logic in expressing the relationship between
generalized and specific categories. In logic, "formal" does not mean white
tie and tails, but rather the *form* or arrangement of items or ideas in a statement, a concept perhaps better understood as the root of the work *formula,*
with its connotation of exactitude in the arrangement of its parts.

Some claim that formal or syllogistic logic is not used in present day biological science but, as Beckner (1959) has indicated, most scientific proposi-
tions and tests represent a compounding of many simple logical units, most of
which are left unsaid but are contained in the taken-for-granted context of
the specialty involved. For example, the Hardy-Weinberg law is a logical hypothesis resting on simpler logical hypotheses involving gene transmission,
meiotic segregation patterns, recombination ratios, and others. We now simply say that a study population has certain genetic characteristics because of
its relation to the Hardy-Weinberg law, without reiterating the many compounding hypotheses on which the law is based. When we use the term "Cretaceous period" for the age of a fossil, we usually don't think of "Cretaceous"
as a compound theory. But in first determining a certain fossil bed to be in
any one geologic period, myriads of logical small hypotheses are invoked, involving faunal or floral assemblages, stratigraphic position, radioactive dating
methods, and so forth.

Inductive-Deductive Reasoning

The form of reasoning used by scientists is the inductive-deductive method.
A distinction is often made between inductive and deductive reasoning. The
differences were explained by William Whewell in the early 19th century
(Butts, 1969). In our actual use of them the two form a peculiarly spiral type
of mental activity as implied by Hull (1967). In the inductive part, individual
data or concepts are marshaled, and from a consideration of them we reach a
general proposition, model, or hypothesis that explains them. For the deductive part, we may deduce from our hypothesis just constructed that certain
results or conditions would follow, and from this, in turn, other results or conditions would follow. This chain of reasoning is continued until we reach a
result or condition that is subject to test, and we make this test. If the test is

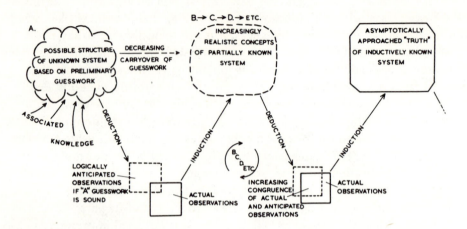

Fig. 4.1 Conceptual schematic illustration of the roles of inductive and deductive reasoning in experimental design. (From A. W. Ghent, "The logic of experimental design in the biological sciences," *Bioscience*, **16**: 17-22, 1966. Reprinted with permission of the American Institute of Biological Sciences.)

not as predicted, but the steps in our deductions were correct, then our hypothesis is false. If the result of the test is as predicted, the original hypothesis is supported. Ghent (1966) has given a flowchart of these steps (Fig. 1), depicting the gradual increase in our understanding as additional observations are added and inferences made.

It is important to realize that a positive result *supports* but does not *prove* the hypothesis, because some other hypothesis or emendation may later be found that is a much more probable explanation of the result.

The above summary of the inductive-deductive method of reasoning is the classical one applied to investigational fields in which experiment is the primary expression of testing. In many areas of systematics, experiment is impossible because no living specimens are available, as is the circumstance with fossils, with species known only from preserved material, and with species that we have not been able to rear in the laboratory. But as Ghent (1966) pointed out, many observations in the field are actually in the nature of experiments. He gave as an example a piece of wood floating down a stream; one could estimate its specific gravity under these circumstances just as well as by placing the same piece of wood in a laboratory tank. In systematics a comparable observation would be that diploid and tetraploid sister species occurred together in the same area, but no triploid individuals were ever observed. This situation would indicate that in the immense laboratory of the field, these two species either did not hybridize, or that for some reason the hybrids were not viable. If large numbers of two distinctive fossil species were found mixed together in the same stratum and no intermediates were observed, the same conclusion would be made. For this reason, the use of "observation" rather than "experiment" in Ghent's figure is of especial interest in systematics.

In this figure, however, the use of the word "guesswork" is unfortunate. The story is told of an attorney trying a case who was challenged by both the judge and the opposing counsel in making a critical distinction between guesswork and estimation. In response he held up a clenched fist and said to the judge, "I have a piece of string in this hand. How long is it?" The judge replied, "I haven't any idea how long it is." To which the attorney responded, "But you could guess?" The judge conceded this to be so. The attorney then held the piece of string taut between his outstretched hands and said, "Now would your honor care to estimate the length of this string?" The obvious point is that in *guesswork* we have no evidence bearing on the answer; our *estimation* is based on evidence. As Ghent pointed out later in the same paper, the "guesswork" in Fig. 1 is based on some associated knowledge or evidence. On this basis it is better termed "estimation"; it is actually the starting induction or hypothesis.

In discussions on the topic of inductive-deductive reasoning, Throckmorton (1972) suggested that, in special reference to systematics, a hypothesis generates questions leading to additional observations, whether experimental or not. Thus if the structures of adult flies lead to the inference of a certain hypothetical phylogeny, we must also ask the question, "How do the characters of the eggs, or the larvae, mesh with this hypothesis?" This question would lead to an observation of these additional characters. Throckmorton made the very important point that if a new hypothesis is needed to explain the added information, it should be made as an inference based on the totality of the old *plus* the new information available, without prejudice as to hypotheses based on only the old information. For example, it was long considered that the fleas were most closely related to the true flies (the insect order Diptera) because in both fleas and flies the larvae are legless, extremely simple, and wormlike. The induction or inference was that *in specialized characters, fleas showed a closer relationship to true flies than to any other insects.* Later (Richards and Richards, 1969) it was demonstrated that the fleas and scorpionflies (the insect order Mecoptera) were the only two insect orders possessing a very peculiar and specialized set of pulverizing teeth in the adult gastric mill (part of the digestive tract). Because the chances were remote that such a complex structure would evolve twice independently, whereas reductions in larval structure have occurred independently in many distantly related orders, it became much more reasonable to consider that the fleas were more closely related to the scorpionflies than to the true flies, and that the lost lar-val legs of the fleas and true flies represented two similar but independent par-allel evolutionary developments. The original hypothesis was then discarded and a new hypothesis formulated: *In specialized characters, fleas show a clos-er geneological relationship to scorpionflies than to any other insects.*

BASIC PREMISES

Of particular importance to scientists are the premises or basic assumptions on which our hypotheses or ideas ultimately rest. Biological science differs in many aspects from the mathematical, physical, or chemical sciences in that its basic concepts or premises may not form as secure a platform on which to build a framework of logical deductions as those in the more exact sciences. The mathematician can rely on basic concepts that cannot be proved, only demonstrated, yet which seem obviously true; for example, that in a finite system the sum of the parts is equal to the whole, that things equal to the same thing are equal to one another, or that two straight lines cannot enclose a space.

In biology the basic premise is often a type that does not lend itself to neat mathematical consequences. We may cite several examples; for example, the premise that all living units have an ontogeny and a phylogeny, or that genetic differences produce physiological difference (but not that physiolog-ical difference is always due to genetic difference). The biologist should also be acutely aware of another premise: he is dealing with a world of change. Thus not only do all living things change from moment to moment and gen-eration to generation, but the environment on which life depends for its exis-tence, and which in part fashions life's form, also changes. This means that, when dealing with living things, the *exact* conditions under which an experi-ment is performed or an observation is made never occurred before and will never occur again.

The biologist is nevertheless in an enviable position. Because many bio-logical problems are eventually resolved into those of chemistry, physics, and mathematics, the basic premises of these fields are therefore available for bio-logical application and ultimately for the construction of biological hypoth-eses or theories.

Although hypotheses are so extremely useful in our thinking, in biology the term *hypothesis* is seldom used in expressing conclusions. If these latter deal with isolated phenomena, we usually write "Based on such-and-such data and certain theoretical considerations, it is apparent that populations *A* and *B* are distinct species." If a hypothesis explains a general pattern of life and has been sufficiently well tested to have a high probability of being correct, we usually refer to it as a theory. Examples include the theory of biological evolution, the theory of evolutionary change by natural selection, the various theories concerning photosynthetic activity, and so on. The sum total of these general theories have predictive values that can be applied to the attempted solution of many new questions and form the theoretical background of much of our thinking.

THE INFERENCE OF HYPOTHESES

We have been speaking blithely about hypotheses and emended hypotheses but have avoided the question: How does one think up a hypothesis? Very simply, a hypothesis is a relationship between different phenomena that someone notices. It might appropriately be called an invention of the mind, because no one has ever been able to decide what inventiveness is or what causes it. Many scientific hypotheses flash into a person's mind when he or she is drying the dishes or painting the garage. They just happen. There are several interesting features about hypotheses that may help explain this situation.

1. From childhood we build one hypothesis after another to attempt to bring relatedness or cohesion into our pattern of sensing and judging.

2. All hypotheses seem to be an emendation of or a substitution for a previous one.

3. We often see evidence not in complete accord with an established hypothesis but reconcile the contradiction by adding an additional assumption to our reasoning rather than change the hypothesis.

4. The neurophysiologists tell us that even when asleep our brain is full of "noise," generated by little electric relays that "like" to look for problems to solve.

5. Perhaps this subconscious problem-seeking activity of the brain somehow solves these problems of contradiction and then somehow signals some part of our consciousness that the deed is done. Perhaps it somehow overrides our preconceived notions of specific points of order in the universe.

The best course of action at the present seems to be to accept the fact that hypotheses just pop into our minds, and let the neurophysiologists eventually tell us how.

Hypotheses are not, however, simply the unbridled activity of the imagination. Each hypothesis is an attempt to explain some formulated question from a set of assembled information. The scientist must assemble information he considers pertinent, and this usually requires considerable knowledge and experience in the field of investigation. From the gathered information he infers an explanation, which is the hypothesis. It is this moment of inference

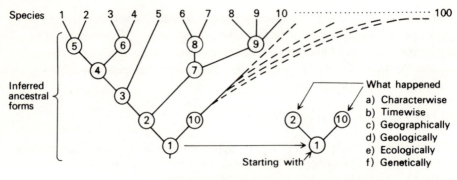

Fig. 4.2 The family tree as a complex model. For explanation, see text.

that is so tricky. Sometimes it is suddenly realized that another observation fits into the scheme and this added bit illuminates the problem. Sometimes the light dawns that a hypothesis being used as a stepping-stone or assumption is wrong; when changed, everything fits.

Because the phenomena involved in systematics are evolutionary events, systematics leans heavily on the hypothetical background of genetics, comparative anatomy, ecology, geology, and various hypotheses of evolutionary theory. This may sound as if systematics was simply evolution. It is true that the two have much in common but they overlap rather than coincide. In systematics the central theme embraces species and their associated lineages, whereas in evolution the major point of focus is biological change in time.

In systematics many working or operational hypotheses are poorly understood and therefore are often in need of change. Notable examples are our hypotheses that certain character stages are the ancestral condition of that character, or that certain land bridges existed at certain times. When any of these are discredited, all hypotheses based on them may also be wrong and must be reexamined.

Because life and its environment are varied in the extreme, so are their interrelationships. To express these complexities, biologists frequently construct models, of which simulated computer models are a good example. Any one of these models is in reality a system of many simple hypotheses, and the model as a whole is dependent on the correctness (always in a probabilistic sense) of its component hypotheses.

Systematics uses a complex model that is often not considered as such. This is the family tree. If intended as an expression of scientific investigation, each fork of a proposed phylogeny is or should be a hypothesis concerning at least what character changes occurred in each branch of the fork. Sometimes each fork is a complex model, in that not only character changes are hypothesized, but in addition temporal, geographic, geotectonic, ecological, and genetic events are added to the hypothetical background. Each explanation of these parameters would (if probable in the light of evidence and deduction) increase our knowledge as to the total circumstances involved in the events associated with any particular fork in the family tree or phylogeny under study. When you consider that a proposed phylogeny of as few as 100 species could include 99 forks (less if some forks had more than two branches), the immensity of such a complex model is readily apparent (Fig. 2).

But as in all other complex models, each individual unit hypothesis of the family tree is an inference based on evidence interpreted according to certain assumptions. New evidence or new ways of looking at old evidence may at any time necessitate a change in any one of the unit hypotheses, resulting in possible changes in other parts of the tree.

Because there is always a chance of demonstrating that any of our premises are either faulty or incomplete, all of our conclusions can, at the most, be hypotheses. Hence, although we seldom say it, we should always realize that the word "if" precedes all of our assertions. Returning to the robin, we should say "If robins always lay blue eggs, this robin will lay a blue egg." This "iffy" nature of our basic premises has a serious consequence for our biological reasoning. Because our simplest basic biological assumptions must be preceded by "if," all subsequent hypotheses resting on them carry this "if" with them. In this light, Popper (1959) insisted that no hypothesis is truly scientific unless it is stated in such a way that it can be proved wrong. Although not all logicians agree with such a rigorous view, none would deny that a hypothesis should be explained clearly enough to invite tests demonstrating its possible falsity.

The emphasis on negation, "if," and the inability to prove hypotheses may give the impression that we can never get anywhere in the pursuit of scientific discovery. This is far from the truth. As more and more evidence and better inductive methods are introduced into our thinking concerning any particular question, we will have to change our conclusions or hypotheses. But as this process continues, successive sets of changes should be smaller and smaller. In systematics, our conclusions as to what is probably a species in this or that group or what is the family tree for this or that group should approach more closely the theoretical ideal of what is actually the case. Turning again to chemistry for an example, hypotheses concerning the structure of the atom have been changing constantly since Lord Rutherford discovered radioactivity at the turn of the century. Before that, the hypothetical atom was a dot with one or more valence arms. Now the hypothetical atom is a miniature but complex universe. In spite of the great advances in our knowledge concerning the structure of atoms, no chemist would think of saying "We now know it all," but all of them would agree that we are getting closer to that point.

Occasionally, new breakthroughs in observation or new experimental opportunities or techniques bring subject matter about which we could only hypothesize into the realm of factual data. For example, before the moon landings, we had no factual knowledge about certain features of its surface. It had been hypothesized from our knowledge of the moon at a distance that all its flat surfaces might be covered with a deep layer of dust into which any landing craft would sink and be buried. After the moon landings it became a known fact that at least part of the moon's surface has no such dust layer. Another example concerned DNA. From X-ray diffraction patterns, it was hypothesized that the strands of the compound were arranged in a spiral. Subsequently the spiral pattern was confirmed factually by electron microscope pictures.

A similar situation has been found in systematics. In certain instances, a plant species has been hypothesized to be a hybrid species resulting from the union of one known parent species with another that was unknown but that presumably possessed certain definite characteristics. Once in a great while this predicted parent species has been discovered, and crosses between the presumed parents produced progeny virtually identical with the hybrid species. This example is not as neat as the moon and DNA examples, but comes extremely close to transferring the case from the hypothetical to the factual realm.

ASSUMPTIONS AND OCCAM'S RAZOR

When trying to reach some inferences from a set of information, we often find that we have several possible solutions that might explain the problem. For example, suppose we find a genus of plants having only two species, one in Oregon and one in Korea. From this we can hypothesize that the ancestral form of these two species dispersed at some past time between the two areas. One question arising from this conclusion is: What was the dispersal route? One person might say, "Across the Bering Straits from Asia to Alaska." Another might say, "No, the route was from Oregon to South America, then to Africa, then to Europe, and from there to Korea." Which is right? Both might be, but on the basis of only the facts given we would take only the first answer seriously.

Why? The answer to this question is *Occam's razor,* attributed to William of Occam (or Ockham), an English Franciscan monk and philosopher of the early 14th century. His great contribution to scientific inquiry was the realization that the greater the number of unwarranted assumptions used in an explanation, the greater is the approach to unbridled imagination, or pure guesswork. He therefore proposed the axiom: The explanation requiring the fewest assumptions is the most likely to be correct. Please note that he said not "is correct," but "is most likely to be correct." Occam must have realized the probabilistic nature of scientific inquiry, even though it was not specifically so stated until a century later by Huygens. He was selling the idea that extra assumptions were *not supported by evidence.* He brought to science essentially the *reductio ad absurdum* principle of Euclid. In other words, if the evidence does not require another assumption, we have no basis to assume one. This principle is also called the *law of logical parsimony.* In our example, the first person assumed only one intercontinental dispersal, the second assumed four.

But one should be extremely careful about the use of Occam's razor. It is meant to guard against assumptions for which there is no evidence. Often evidence that negates a previous conclusion based on Occam's razor will turn up. For example, the tsetse fly *Glossina,* vector of the organism causing sleeping sickness, now occurs only in Africa, and Africa was earlier considered both its present and ancestral home. If earlier in the century someone had said "It formerly occurred in the New World," he would have been voted out, and rightly so, for making assumptions without evidence. Later, fossils of *Glossina* were discovered in the Florissant shales of Colorado. Because of this and many similar examples, it is advisable always to consider various alternatives that might explain a situation. This method of approach might lead to assembling unusual sets of information which might yield scientific surprises.

Criticism is frequently leveled at biological scientists of all types that they are extending their hypotheses far beyond the point permitted by their data and are indulging in "speculation" or "guesswork." Probing into the dark, however, is one of the most important features of the scientific method. It can be dignified by the terms *abductive reasoning* or *inference* (Hanson, 1958; Platt, 1964), but in plain language it is simply *stretching the mind.* Francis Bacon may have been the first to point out the importance of this activity; certainly he asserted it vigorously. Inference can take almost any form, from realizing inductively the general pattern contained in a set of data or deducing unrealized consequences of a hypothesis. At its best it is the flash of realization whereby we suddenly see a possible answer to a question that has been nagging at our minds. These engendered possibilities are then subject to test as would be a hypothesis, and those that survive testing become part of our body of theory.

THE VALUE OF ARGUMENT

One of the most stultifying influences in biology (as in any science) is the suppression of differences of opinion within a field of investigation. There are both logical and philosophical reasons for such differences. If we bring together *all* evidence concerning a problem, we must admit that some of it may be *contradictory* evidence; to explain this may require two or more alternative hypotheses each with a relatively low probability. This is a logical basis for differences of opinion. The philosophical reason is equally simple. All scientists are humans, and no two have exactly the same backgrounds of personal experience against which to compare alternative hypotheses. It is to be expected that each person may consider different pieces of evidence as being more significant concerning any one problem, resulting in honest differences of opinion. But certainly the expression of these differences, *if clearly explained,* is one of the most powerful tools in stimulating greater search for new evidence or more careful examination of hypotheses and deductions.

We need to emphasize one point that may not have been stressed sufficiently. Throughout any investigation, curiosity should be kept in the forefront of one's thinking. Curiosity is at the root of opening new avenues of looking at things and stimulating suggestions of new phenomena to explore.

It may seem that the demands of logic and the scientific method take the adventure out of scientific discovery. This is not so. These logical precepts simply demand that in any investigation we must

1. Outline our questions unambiguously.
2. Collect all pertinent evidence possible.
3. Understand the basic premises and theory governing or applicable to the scientific area under study.
4. Interpret the evidence rigidly (in a logical sense) and in accord with the body of theory.
5. State the conclusion or inference clearly.

Looking back on some of our past questionings, it seems almost inconceivable that man, standing on this little flyspeck called the earth, could gaze into the heavens and finally figure out that his flyspeck revolved around the sun, that the solar system and millions of other stars formed a giant wheeling galaxy, and that beyond it were millions of other equally large galaxies. The process of discovery was a long one, full of false starts, some good deductions, and a continual testing of ideas both new and old. Progress was stymied at times by lack of technical advances in equipment or lack of physical or mathematical theory. Skeptics periodically predicted failure to further understanding. But in spite of all difficulties, attempts to solve questions were never abandoned.

No less amazing in retrospect has been the discovery of the chemical elements, then their bonding properties, and now an understanding of chemical compounds that 200 years ago would have seemed absurd to contemplate. Yet a corps of early chemists must have glimpsed at least some of this "absurdity," had faith in man's ability to comprehend, and persistently pursued every clue.

Today's systematists are in an enviable position. They know enough about the myriad species on the earth to realize the tremendous potential in increased understanding that is within the grasp of the inquiring reasearcher. Great strides being made in related fields such as functional morphology, physiology, cytogenetics, biochemistry, ecology, and economic biology are contributing a deluge of information concerning species characteristics that will be of use in systematics. Systematists themselves have new tools of discovery such as electrophoresis, new types of light microscopes, various electron microscopes, to say nothing of unparalleled opportunities for collecting specimens from previously inaccessible areas. In the next decades they can expect a fantastic increase in new evidence available in biosystematics. They have a scientific method to follow that will keep them headed in the direction of more and more probable conclusions.

Systematists are just at the beginning of one of biology's greatest adventures in achieving a greater understanding of life.

5

The Biological
Processes of Speciation

To obtain results of the greatest significance in studies of ecology, evolution, and most fields of comparative biology, it is essential to know how many species are present in the study area and how to tell them apart. For this reason an understanding of species and the concepts they represent is an important aspect of biology.

In essence, species are the different kinds of living things. The term *species* has long been almost a magic word in biology. It is one word that everyone knows. We argue about what it means or should mean, abuse it, proclaim its deficiencies, and offer substitute terms in place of it, but the term *species* survives. As Dr. W. R. Richards once said, "You can't define it and you can't get along without it."

Why this persistence? It is undoubtedly because, in some guise such as "kind" or "sort," the term is as old as primeval man. Primeval man may have had no more conscious sense of classifying the living things in his environment than do birds or nonhuman mammals, but even this taxonomic perception appears to be more considerable than often thought. Certainly by the time man evolved into well-established hunting tribes, he was a discriminating and observant naturalist. Highly nomadic and naturalistic groups still in existence give good evidence on this point. In one district having 102 species of birds, Eskimos recognized 98 correctly, grouped the other four into two kinds, and had a distinctive name for all 100 taxa they recognized, even though only a handful of the species were of importance in the Eskimo economy (Irving, 1953). Distinguishing the noneconomic portion of the biota represents man's typical curiosity concerning his surroundings; naming them represents man's desire to talk about them. It is difficult to know when man developed a biological sense of species, but certainly he had this sense by the time he became a plant and animal breeder in the earliest agricultural economies.

Early historic peoples, especially the Greeks, used species extensively in their observations. Later, through the period of the Roman Empire and the following Dark Ages of Europe, the term species was undoubtedly recognized as a general concept but nothing was written about either it or natural history.

When Europe emerged from the Dark Ages, one of the first concepts to be put on a basis that we would call scientific was that of the species. This happened first in the mid-16th century by the Englishman Wooton, the Swiss Conrad Gesner, and the Italian Ulisse Aldrovandi, all of whom published highly accurate accounts and illustrations of the plants and animals occurring in their respective countries. In the 17th century the English naturalist John Ray produced the first rudiments of our hierarchal classification and the first good definition of the species as a reproducing unit. This definition followed practices used previously by plant and animal breeders, but the empirical methods of this applied group were held in low esteem in Ray's time and were not used as scientific evidence until Darwin did so two centuries later.

After Linnaeus provided a simple system for naming and identifying species in the mid-18th century, more and more kinds of plants and animals were described and diverse types of biological and geographic information were associated with them, using the species name as the key card in the storage of information. By the end of the 19th century the species, very much as we recognize it today, had been accepted as the basic unit expressing the different kinds of life on the earth. An enormous amount of man's biological information was and is filed under the names of these kinds or species. No wonder the species is considered practically a sacred scientific category!

By or soon after the turn of the 20th century, difficulties in defining, equating, and identifying species began to accumulate. To name a few of the difficulties, intermediate populations were discovered between species previously considered distinct, asexual and apomictic species were recognized, and cryptic species and the role of ploidy were brought to light. To circumvent these difficulties, several definitions of species were proposed. The first was one based on difference in morphology. The discovery of cryptic species undermined this morphological definition. Next came a genetic definition, stressing lack of hybridization as a species criterion. This was favored by zoologists for many years but not by the botanists. Now some investigators realize that it is not a satisfactory criterion for animals either. These definitions and associated arguments were concerned almost entirely with bisexually reproducing species. Asexual and apomictic types were largely ignored. A separate problem arose with certain aspects of naming those fossil species presumably representing different points on the same lineage.

The solution to the dilemma is simple. There are several types of species, and concepts that apply to one type do not necessarily apply to the others. We are not dealing with a species problem, but with species problems. Each must be considered separately.

TWO DIMENSIONS OF SPECIATION

Speciation may be defined as any process by which one (or more) species gives rise to another (or other) species. At this point we must work on a simple operational definition of species requiring only that two entities or populations are sufficiently different that we want to give them different names in order to discourse about them intelligently and unambiguously.

Speciation occurs by either of two distinct processes: by changes in time within the same lineage, or by the production of additional lineages.

Evolutionary Change in a Lineage

As the successive generations of a lineage or phylogenetic line progress through time, the succeeding populations of the lineage change to a greater or lesser degree. If we imagine a lineage that existed 3 million years ago, at the dawn of the ice age, with annual production of young, it is almost certain that the genetic constitution of a class of any one year was different from that of the preceding or succeeding year. The result would be a change with time in the characteristics of the populations comprising the lineage, or in different geographic populations of the lineage.

Disregarding geographic races, we can surmise that successive populations of the same lineage would change with time as in Fig. 1. The character modes of each generation would differ from those of its predecesor in a somewhat irregular fashion, but such that each generation might be substantially different. In the example, from Time 1 to Time 7 lineage *A* changed to such an extent that all individuals existing at Time 1 were diagnostically different from all individuals existing at Time 7. One could therefore say that lineage *A* had evolved into a *different kind* of organism, that is, into a different *species,* between Times 1 and 7. To express observations about these two species, each would need to be given a distinctive name.

In our historic observation of the populations representing ongoing lineages (which we call species), we frequently see little or no difference over periods of several hundred years. For example, herbarium specimens of Nearctic *Quercus* species (oaks) over 200 years old look exactly like members of the same species collected today. Fabrician type specimens of bumble bees (*Bombus*) are virtually identical with specimens collected today. Examples such as these give the impression that character change in a phylogenetic lineage is extremely slow. On the other hand, since its introduction into North America in the mid-19th century, the English house sparrow *Passer domesticus* has differentiated into at least a dozen fairly-well marked Nearctic geographic forms equivalent to many named races of other species (Selander and Johnston, 1967). A goodly number of European insects inadvertently introduced into North America have apparently evolved new physiological charac-

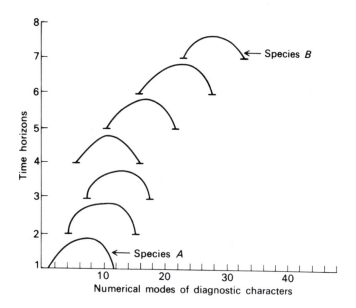

Fig. 5.1 Diagrammatic representation of possible changing character modes in successive populations of a phylogenetic line or lineage starting with species *A* and progressing to species *B*.

teristics in a matter of only 10 or 15 years (Elton, 1958). Thus, whereas our average records of change may seem slow, other records demonstrate a rapid rate of change. Not measured, however, is the probable low rate of change of the homeland populations of these introduced species.

Although the interjection into the argument of this latter point may seem confusing, it is of paramount interest concerning a much debated point about evolutionary changes in lineages. The stay-at-homes, remaining in relatively unchanging ecological conditions, tend to change slowly. In populations or species subject to altered or altering ecological conditions, change is potentially much more rapid. As a consequence we can conclude that change in a phylogenetic lineage is *not a function of time alone,* but of *time combined with changes in values of natural selection.*

Thus, in Fig. 1, under relatively constant ecological conditions, the absolute time interval between Times 1 and 8 might be over 100 million years; under conditions of changing ecological conditions, it might be as low as 50 thousand to 100 thousand years, or even less.

Changes that have occurred in individual lineages are documented by fossil species found at different time periods. Each distinctive type or species

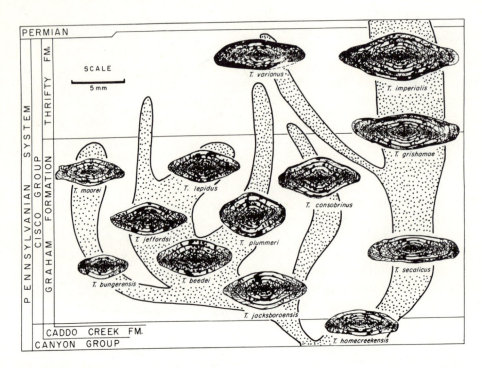

Fig. 5.2 Phylogeny of common species of the fusulinid protozoan genus *Triticites* from Upper Pennsylvanian strata of Texas. Horizontal lines represent stratigraphic subdivisions. (From C. A. Ross.)

presumably representing a single phylogenetic line is termed a *sequential species.* For example, in the Fusulina (the giant Foraminifera of the Paleozoic era), three species of the genus *Triticites, secalinus, grishamae,* and *imperialis,* form a series of sequential species arising from *T. homecreekensis* (Fig. 2).

Types of Lineage Change

Lineages may change genetically through time by either intrinsic or extrinsic change, or both. In *intrinsic change* all the genetic change arises within the species comprising the ongoing lineage. Most of the change would normally be in the form of recombinations of alleles brought about by crossing-over and chromosomal rearrangements. Mutations would also contribute to genetic change, but it is thought that these occur much less frequently than gene recombination. Loss of either certain genes or gene combinations would also contribute to change.

In *extrinsic change,* made famous by the botanist Edgar Anderson (1949) under the name *introgressive hybridization,* the change occurs following a certain kind of hybridization between different species representing distinctive lineages. If the hybridization between two lineages or species, were massive and continuous, and the hybrid individuals were successful and crossed readily with each other and both parents, the two species would eventually merge into a single species. We would conclude that the original parents had never reached the state of being distinctive species.

If the hybridization is below the merging value, the hybrids are usually few or intersterile, but they will often backcross to a small degree with one or both parent species. When this backcrossing occurs, it introduces a small amount of the genetic material of one species into the genetic make-up of the other. If the new genes bestow a positive selective value on their new carrier, they will become incorporated into the gene pool of that carrier. As a result, the two parental species continue as separate lineages but at least one of them has been changed by the addition of some genetic material from the other.

During the course of its history, a lineage could very well be changed genetically by introgressive hybridization with a succession of different species. There is also the possibility that a lineage might have introgressive relationships with more than one species simultaneously, as Sauer (1957) demonstrated with the amaranth pigweeds. Compound introgressive influences of this type are readily (but not easily) detectable if they are happening at the moment and are therefore subject to experimental test, but they would be difficult to ascertain on the basis of fossil evidence collected at different time intervals.

The role of introgressive hybridization in causing genetic change in an evolving lineage has been recognized commonly in plants (Anderson, 1949; Stebbins, 1950; Heiser, 1949, 1951), but in only a few instances has the phenomenon been recognized in animals (Sailer, 1954; Miller, 1955; Ross, 1958; Mecham, 1960; Niethammer, 1969; Remington, 1968). A situation surely resulting in introgression has been described in detail for the golden-winged and blue-winged warblers, *Vermivora chrysoptera* and *pinus,* of eastern North America (Parkes, 1951). The two species have a broad area of overlap from southern New England to northern Illinois and southern Minnesota. Here they hybridize, in some areas freely, but the hybrids almost invariably backcross with the parental types they most resemble. Field observations indicate that both the hybrids and the backcross progeny are vigorous and viable. The hybridizing pattern keeps the two parental types highly distinct on a modal basis but certainly results in constant introgression between the two species. Unfortunately, no exact figures on amount of introgression are available. Undoubtedly more instances of introgression will be discovered in all groups of organisms.

Anderson postulated that, in introgressive hybridization, integrated blocks of genes would become transferred from one species to another. Such transfers would result in the introgressed species having a mosaic inheritance of certain characters. Ross (1958) found that in the leafhopper genus *Erythroneura,* species exhibiting strong evidence of introgression did indeed show a mozaic pattern of parental character states. In natural hybrids be- tween *Senecio smallii* and *S. tomentosus,* Chapman and Jones (1971) found that several characters were inherited in a mosaic pattern, although the major- ity were expressed in the hybrids as intermediates between the parental types. (Fig. 3). The presence of mosaics of parental character states presumably multigenic in nature is most helpful in detecting probable introgression. A mozaic pattern of inheritance is not, however, a necessary requisite of intro- gression. As is obvious from the Chapman and Jones' study, introgression may result also in character states intermediate between those of the two par- ents. The emphasis on mozaic inheritance stems from the fact that in nonex- perimental studies the mozaic patterns are an unusually good criterion for de- tecting introgression.

In his analysis of introgression, Anderson (1949) associated the phenom- enon with the occurrence of "hybrid habitats." For example, if one species grew in loam soil and its related species on rocky slopes, the hybrid individu- als would occur in areas intermediate ecologically between those occupied by the two parents, that is, in gravelly areas with gravelly soil. The rationale was that the hybrids would be outcompeted by either parent species in their re- spective ecological setting; only those hybrids could survive that occurred in a habitat in which they had an advantage, or were at least not at a disadvantage, with regard to parental competition. Anderson had several examples of the occurrence of occasional hybrids in such "hybrid" environments.

Other observed examples of introgression suggest an association with an added circumstance. In Illinois, abundant hybrid swarms of *Quercus marilan- dica* and *ellipsoidalis* occur primarily on sand areas which are different from, but not intermediate between, the typical habitats of the two parents. In the central United States, hybridization between species of sunfishes is encountered most frequently in artificial ponds that are not intermediates between the typ- ical habitats of the parent species. Mecham (1960) encountered introgression in frogs in similar artificial situations. All three examples plus several cited by Anderson (1949), especially the introgression observed in *Iris,* involve newly created or disturbed habitats in which either there is little competition or the normal ecological cues triggering sexual behavior are changed. Such circum- stances appear to favor interspecific hybridization and, in turn, introgression.

In summary, evolutionary changes in a lineage reflect an amalgam of both intrinsic and extrinsic events. If these are occurring at the present, it is often possible to learn much about the mechanics of these events. Techniques for deciphering past happenings of this type are largely unknown, but a realiza- tion of their probable existence should lead eventually to better means of de- tecting and understanding them.

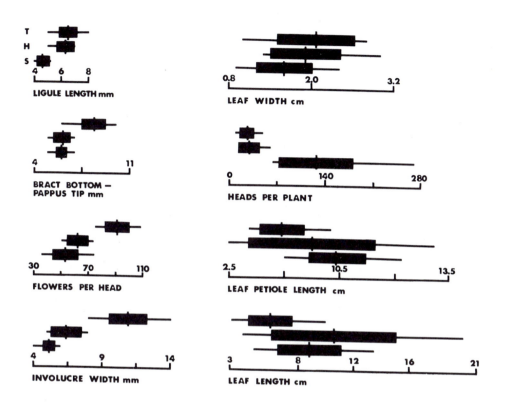

Fig. 5.3 Mean, range, and standard deviation of selected characteristics of *Senecio tomentosum* (T), *S. smallii* (S) and their hybrids (H). Note the mosaic inheritance of ligule length, bract bottom–pappus tip, and heads per plant. (From G. C. Chapman and S. B. Jones, Jr., "Hybridization between *Senecio smallii* and *S. tomentosus* (Compositae) on the granitic flatrocks of the southeastern United States," *Brittonia*, 23: 209-216, 1971. Reprinted with permission of The New York Botanical Garden and the authors.)

Naming sequential species must be done on an arbitrary basis. In a sequence of species only the upper member could be an extant species; all others would be based on fossils. There is no uniform criterion for deciding how different successive fossil populations should be for them to be designated as different species. Systematists working with a particular group usually have a set of judgments based on amount of difference and amount of character overlap between successive populations for making these taxonomic decisions.

The Production of Additional Lineages

When one considers that initially life consisted in perhaps only a single lineage, yet today comprises at least 2 million species each representing an evolving distinctive lineage, it is immediately apparent that one or more mechanisms exist by which additional lineages come into being. Such additional lineages would result in an increase in the number of species occurring at any one time, because species are time-transects of an evolving lineage.

The phenomenon of the production of additional lineages, in other words, an increase in the number of phylogenetic lines, has been given many names. Haekel (1866) termed it *phylogenesis*. Others have given it the name *clado-genesis* (the birth of branches) and *phyletic branching* (Simpson, 1944). I prefer the unabbreviated form of the latter, *phylogenetic branching*.

Investigations to date have demonstrated many mechanisms by which this increase in number of species or lineages can occur. These mechanisms involve bisexual, unisexual, and asexual species.

From the standpoint of speciation, type of sexuality in itself is not the important criterion. The fundamental division is whether the species is panmictic or apomictic.

In *panmictic species* there is a mixing of gametes between different individuals at some stage in the life history. Species in which each sex is produced by a different individual (*dioecious*, having two homes) are always panmictic. This includes many plants such as the ginkgo and hollies and most metazoan animals. Species in which the two sexes are produced by the same individual (*monoecious*, having one home; *hermaphroditic*) are panmictic if some of the progeny are the results of cross-fertilization between different parent individuals. Examples include the lumbricid worms, most flowering plants having stamens and ovules in the same flower, and plants such as pines and many sedges in which each catkin or spike is either male or female but both types are borne on the same plant. In many hermaphroditic individuals, gametes of the two sexes mature at different times, thus ensuring cross-fertilization. The result of panmixia is a lineage having a reticulate structure (Fig. 4a) providing a maximum of genetic recombination within the population.

In *apomictic* species there is no mixing of gametes between different individuals. Most apomictic species are unisexual, in that only ova are produced. Examples include the dandelion and many parthenogenetic insects that produce no males. Other apomicts reproduce completely asexually by budding or fission, and have no functional sexual stage in any part of the life history. Thus certain triploid lilies have well-formed flowers but reproduce only by bulb offsets, the flowers being sterile. Often asexual organisms never produce sexual stages, such as the worm *Enchytraeus fragmentosus* in which the body fragments periodically and each piece grows into a complete worm. The result of apomixis is a clonal type of lineage affording no opportunity for genetic recombination between different members of the population (Fig. 4b).

It seems certain that throughout all groups of organisms the long-continuing lineages have been panmictic to at least some degree. Evidence so far un-

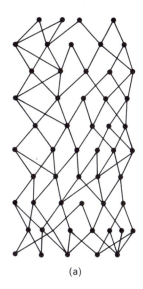 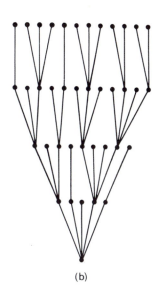

(a)   (b)

Fig. 5.4 Diagrammatic representation of (a) panmictic and (b) apomictic populations. Dots represent individuals, connecting lines their parentage.

earthed shows that apomictic species have arisen from time to time from panmictic species but have not become the basis for long-lasting phylogenetic lines. A few apomictic species have evolved into a considerable number of daughter species or short sequences of species, but compared with panmictic lineages they span only a few moments of geologic time.

Mechanisms for the production of additional lineages are therefore discussed first for panmictic species and after that for apomicts.

MECHANISMS IN PANMICTIC LINEAGES

To date we have evidence for four mechanisms producing additional panmictic lineages:

1. geographic isolation,
2. hybridization,
3. ecological isolation,
4. polyploidy.

Speciation by Geographic Isolation

It was noticed soon after the publication of Darwin's *Origin of Species* that species living in the same area tend to be readily differentiated whereas species occupying separate but neighboring areas tend to be extremely similar. This circumstance led Wagner (1868, 1889), Romanes (1897), and Jordon (1905) to propose that additional species have arisen through the geographic isolation of segments of a parental species. Early evidence was drawn chiefly from animal groups. In 1905 Abrams reported that botanical evidence also supported this view.

In the ensuing years a tremendous amount of geographic evidence has accumulated indicating the important role played by geographic isolation in the evolution of new species. Excellent reviews have been published by Mayr (1942, 1963) for animals, Stebbins (1950), Grant (1963, 1971), and Briggs and Walters (1969) for plants. This evidence falls into four major categories: species having a disjunct range; sister species occupying different ranges (allopatric); sister species whose ranges overlap broadly (sympatric); and sister species whose ranges abut (parapatric).

Disjunct Ranges

One is inclined to think of the range of a species as a neat geographic area, a belief bolstered by pink areas on a map or ovals or ellipses laid out in a generalized or semidiagrammatic form. When ranges are examined in detail, such a simple or idealized pattern seldom results. Two circumstances are responsible for this revised viewpoint. First, the range of a species may differ from year to year in a manner that is entirely unpredictable. An example is the North American high plains grasshopper *Dissosteira longipennis* (Fig. 5). In its years of greatest abundance, the area in which this grasshopper bred was many times larger than that to which it presently extends (Wakeland, 1958). This and other short-term changes in species ranges are paralleled by other long-term changes in range. Excellent examples are afforded by the chaparral and woodland vegetation of the Southwest and Mexico. In mid-Cenozoic time, these or related species also occurred considerably to the north and east of their present restricted distribution in California and Mexico (Fig. 6) (Axelrod, 1958).

From these and many other examples it is evident that the geographic ranges of successive generations of a single pnylogenetic line are as dynamic or changing as is the genetic constitution of that lineage combined with the climatic oscillations of its range. We would therefore expect to find in today's biota, which is a cross-section of all persisting lineages, every conceivable type of species range. And, in fact, we do. In the plants, the bracken fern *Pteris aquilina* has a range encompassing all the temperate and boreal areas of the Northern Hemisphere; at the other extreme, the ironweed *Vernonia lettermanni* is known from only a minuscule area along the banks of the Ouachita River in Arkansas and Oklahoma. In the animals, the puma *Felis concolor* extends from northern Canada through Central America to southern South America; many cave insects in North America are each known only from a single locality.

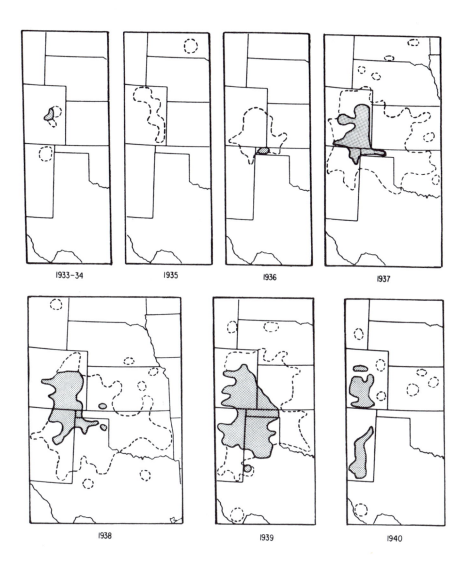

Fig. 5.5 Oscillations of the range of the American high plains grasshopper *Dissosteira longipennis* between 1933 and 1940. The shaded area is the breeding range; the areas shown by dotted lines are areas of adult dispersal from that year's breeding range. (Modified from Wakeland; from H. H. Ross, *A Synthesis of Evolutionary Theory*, © 1962. Reprinted by permission of Prentice-Hall, Inc., Englewood Cliffs, N. J.)

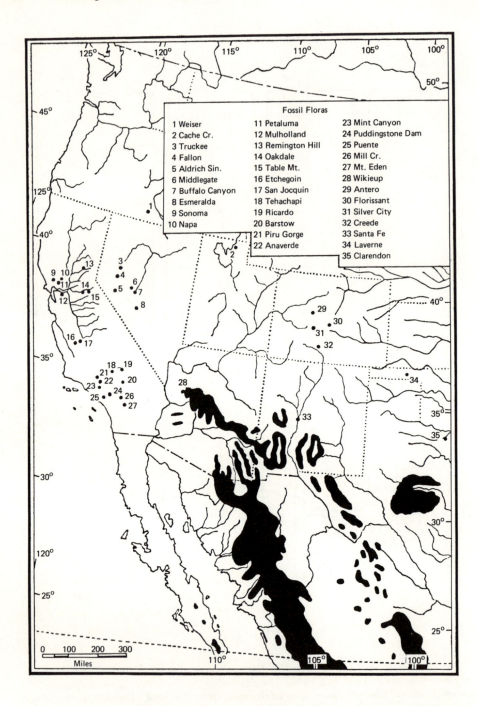

Fossil Floras

1 Weiser	11 Petaluma	23 Mint Canyon
2 Cache Cr.	12 Mulholland	24 Puddingstone Dam
3 Truckee	13 Remington Hill	25 Puente
4 Fallon	14 Oakdale	26 Mill Cr.
5 Aldrich Sin.	15 Table Mt.	27 Mt. Eden
6 Middlegate	16 Etchegoin	28 Wikieup
7 Buffalo Canyon	17 San Jocquin	29 Antero
8 Esmeralda	18 Tehachapi	30 Florissant
9 Sonoma	19 Ricardo	31 Silver City
10 Napa	20 Barstow	32 Creede
	21 Piru Gorge	33 Santa Fe
	22 Anaverde	34 Laverne
		35 Clarendon

Fig. 5.6 Occurrence of present Sierra Madrean woodland (black) and records of earlier r
lated Cenozoic vegetation (numbered dots). (From D. I. Axelrod, 1958)

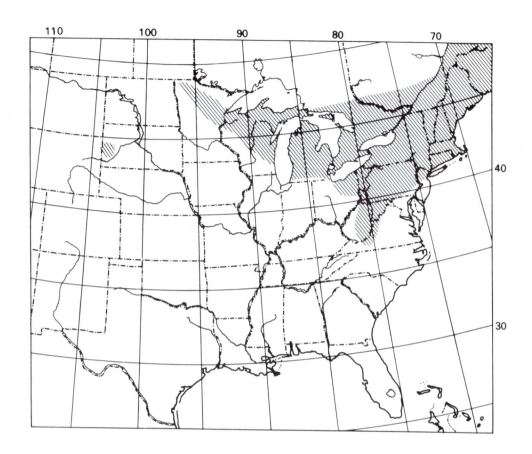

Fig. 5.7 Range of the smooth green snake *Opheodrys vernalis* showing the small isolated population (at left) in the Black Hills of South Dakota. Prepared from information kindly furnished by Dr. P. W. Smith.

Most ranges of any great extent have disjunct populations, that is, populations which occur at some distance from the main body of the species range. If the species is highly vagile (*vagility* is the ability to disperse), disjunct populations indicate only the colonization or recolonization of an isolated area by repeated dispersals of individuals from the main range over or across areas in which the species cannot maintain itself. Such dispersals may be achieved via wind, flood, or fortuitous peregrination.

If the species has a low vagility, then disjunct ranges involving long inter-colony distances indicate the existence in the past of dispersal routes no longer in existence. The green snake *Opheodrys vernalis,* for example, has a main

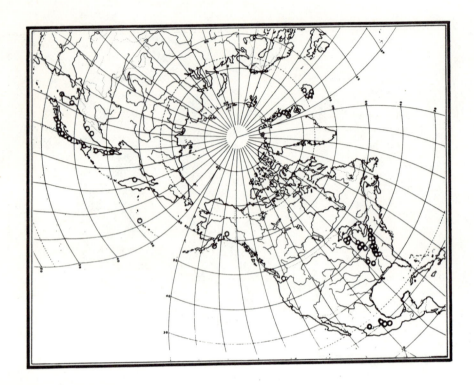

Fig. 5.8 Distribution of the sword moss *Bryoxiphium norvegicum,* showing the wide separation of its populations. (Modified from A. Löve and D. Löve, 1953.)

range in the northern and eastern United States and adjacent Canada, with an isolated disjunct range in the Black Hills of South Dakota (Fig. 7). The species has a vagility so low that the Black Hills population is not explainable as colonization between the two present range sections. We are therefore led to postulate that in some past time ecological conditions between the disjunct areas were different from those of the present and provided habitats in the intervening areas in which *O. vernalis* could exist. A comparable example is the purple skunk cabbage *Symplocarpus foetidus,* having disjunct populations in China and eastern North America.

Some of the more unusual disjunct ranges are as yet open to hot debate. The sword moss *Bryoxiphium norvegicum* has disjunct populations over most of the Northern Hemisphere (Fig. 8), but it is a matter of argument whether the disjunct populations are due to unusual vagility through wind-borne spores or to past range extensions due to changed ecological conditions.

The primary evidence of disjunct ranges as they bear on phylogenetic branching comes from examples involving species of low vagility and relatively

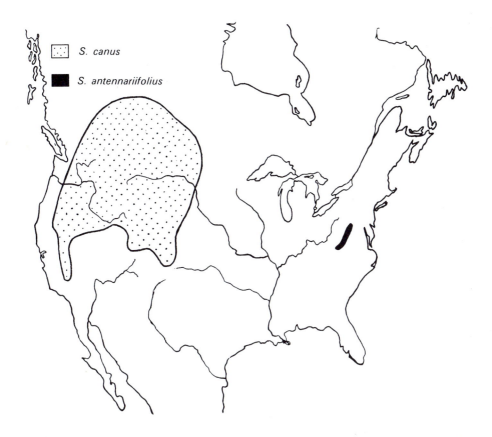

Fig. 5.9 Distribution of two closely related species, *Senecio canus* in the West and *S. antennariifolius* of the Appalachian area. (From G. L. Stebbins, Jr.)

large distances of disjunction such as are provided by the green snake and the purple skunk cabbage. Hundreds of other equally appropriate examples include species of marine crustaceans and mollusks, aquatic and terrestrial insects, vertebrates, and plants.

Allopatric Sister Species

Paralleling the low-vagility species, each having a disjunct range, are multitudinous instances of sister species each occurring in a different area. In two sister species of the plant genus *Senecio*, *S. canus* occurs only in the West, *S. antennariifolius* only in the East (Fig 9). In a pair of sister species of Nearctic stoneflies, *Allocapnia frisoni* occurs only in the Appalachian region, *A. pel-*

toides in the more western Ouachita range. In the caddisfly genus *Wormaldia,* the species *W. kisoensis* occurs in Japan, its closest relative *W. mohri* only in eastern North America.

Most famous of allopatric species are Darwin's finches, a distinct group of Fringillidae restricted to the Galapagos Islands. Of the 14 species, several occur on only one island. Myriads of pairs of related species, each occurring on a neighboring island, have been recorded in many groups of both plants and animals.

Sympatric Species

The third category of species distributions includes sister species whose ranges have a large area of overlap, a common condition known as *sympatry.* In the plants, excellent examples are *Gilia latiflora* and *G. ochroleuca* whose ranges are sympatric but which are incapable of interbreeding (Grant, 1963). Bird examples include species pairs in the American wood warblers of the family Parulidae, especially in the genera *Dendroica* and *Vermivora.* Innumerable examples of sympatric sister species occur in the insects, including many examples each in the leafhoppers, crickets, mosquitoes, caddisflies, and stoneflies.

In both plants and animals, a great amount of crossbreeding has demonstrated that sympatric sister species normally do not interbreed, whereas, if brought together artificially, allopatric sister species frequently hybridize to some extent and often freely. Much evidence for plants is summarized by Stebbins (1950); for *Drosophila,* by Patterson and Stone (1952). On the basis of characters used for species discrimination in their respective groups, the allopatric species pairs were as distinct, or more so, than sympatric species pairs.

Parapatric Species

This category includes sister species whose ranges abut with at most a narrow area of overlap. Most of the recorded parapatric ranges are due to competition combined with critical ecological boundaries. Certain plants in the western United States are restricted to serpentine soils, whereas related species occur on other soils. It is presumed that the species on serpentine soils cannot compete with their relatives on other soils but can on serpentine, resulting in a sharp break between the species at the edge of serpentine areas. Of two wheat stem sawflies of the genus *Cephus* introduced into eastern North America (Udine, 1941), one always outcompetes the other if they occur together, but the loser can live further south, resulting in a sharply defined edge for the two ranges (Fig. 10). In most instances where parapatry is maintained by competition, the related species involved are not newly evolved sister species, hence they represent a historic ecological interaction of considerable duration, possibly following an earlier stage of being two sympatric species.

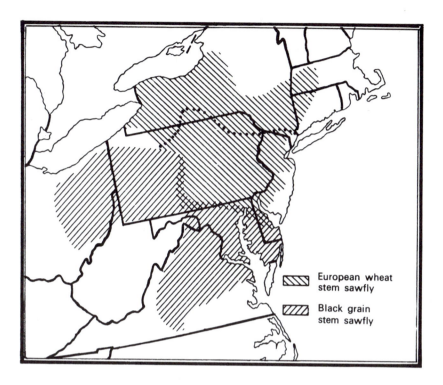

Fig. 5.10 Distribution of the black grain stem sawfly (to the south) and the European wheat stem sawfly (to the north) in the United States in 1937. The dotted line marks the former northern extension of the black grain stem sawfly before the European wheat stem sawfly became established in North America. (From H. H. Ross, *A Synthesis of Evolutionary Theory*, © 1962. Reprinted by permission of Prentice-Hall, Inc., Englewood Cliffs, N. J.

A different kind of parapatric distribution has been described by White (1956) and White et al. (1964) for the south Australian grasshopper *Moraba viatica*. Two forms of it are known, an eastern 17-chromosome form and a western 19-chromosome form. Members of the two intermate readily, but the progeny of hybrid crosses have reduced fecundity. Both forms are wingless and sluggish, moving probably only a short distance from their point of hatching in their lifetime. White found that the area of overlap between these two was only about 1 km, and that few hybrid collections were encountered in it. He concluded that the sharp geographic division between these forms was a result of random intermating combined with low hybrid fecundity and extremely low vagility. He originally called the two forms races, but in an evolutionary sense they are species. To this particular type of evolutionary situation he applied the term *stasipatric speciation* (White, 1968).

These circumstances of species distribution and interbreeding character-istics led to the idea that disjunction of ranges, allopatric differentiation into species, and the subsequent sympatric congregation of the sister species were sequential steps in a process by which one species became two or more sister species through geographic isolation.

To be completely tenable, this hypothesis needed two mechanisms, one providing for the geographic division of a species and the later reunion of its daughter species, and another mechanism by which reproductive isolation would evolve if these newly formed species were reunited.

Geographic range division. The first mechanism was provided by geology and climatology. The crust of the earth is dynamic, producing constant change in both topography and climate. The topographic changes include island and mountain formation and the rise and fall of land bridges. The climatic changes include temperature and precipitation, the former being the only one affecting oceanic life. It has been shown that many geologic events have been reversible, for example, the Pleistocene Ice Ages, and such reversible or oscillating cycles could provide the range-division, range-reunion effect needed for the foregoing speciation hypothesis.

Reproductive isolation. The second needed mechanism was one providing re-productive isolation between evolving sister entities whose ranges became sym-patric. The occurrence of reproductive isolation between even extremely closely related sympatric species has been known for many years. Originally this isolation was considered to be due simply to genetic differences that pro-duced sterile or inviable offspring. When the mechanisms producing this iso-lation in animals were studied, another factor came to light, that of mating be-havior. Distinctive courtship patterns specific to a single species were found in mice (Blair and Howard, 1944), fish (Gordon, 1947), field crickets (Fulton, 1952), *Drosophila* (Spieth, 1952), and many other groups. The distinctive male coloring of bird species and various insects are undoubtedly associated with courtship patterns. The importance of these discoveries lay in showing that related sympatric species are *sexually isolated* from each other, a form of isolation that prevents matings between the species and thereby increases the reproductive potential of each species. This is the mechanism of *reducing gametic wastage.*

Studies in animal-pollinated plants verified this hypothesis. Many distinct plant species were each discovered to be pollinated by a different insect spe-cies, for example, *Aquilegia pubescens* and *formosa.* When artificially crossed, these two species were found to be fully compatible genetically; however no or few hybrids were found in nature. The insect pollinators obviously pro-vided sexual isolation.

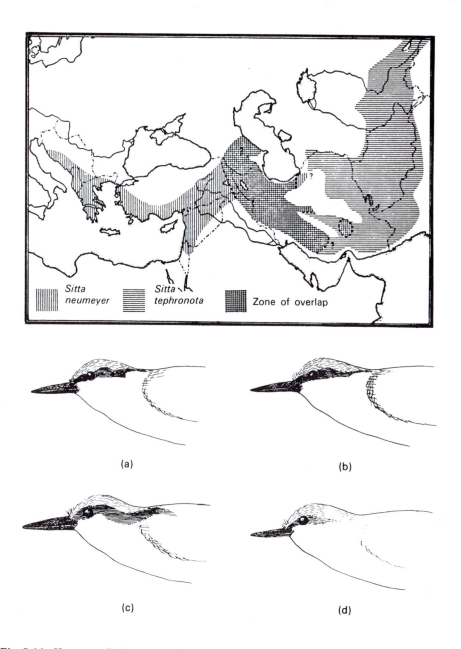

(a)

(b)

(c)

(d)

Fig. 5.11 Character displacement of the beak structure and head pattern in rock nuthatches of the genus *Sitta*. Above, ranges of the two species showing the area of overlap. Below, heads and beaks of the same two species. (a) *S. tephronota* from Ferghana; (b) *S. neumeyer* from Dalmatia; (c) and (d) *S. tephronota* and *S. neumeyer*, respectively, from western Iran where they occur together. (Adapted from Vaurie, in H. H. Ross, *A Synthesis of Evolutionary Theory,* © 1962. Reprinted by permission of Prentice-Hall, Inc., Englewood Cliffs, N. J.)

Stalker (1953) demonstrated that many distinctive but related *Drosophila* species with allopatric ranges will hybridize freely if brought together
artificially, even though the hybrids are sterile or inviable. It was therefore
apparent that these species pairs had evolved genetic differences to the point
of possessing *genetic incompatibility* but had not evolved the differential behavior patterns of their sympatric counterparts; in other words, had not
evolved *sexual isolation.*

These circumstances led to the hypothesis that genetic incompatibility
did not automatically confer or result in sexual isolation, but rather that sexual isolation could arise independently through the action of natural selection.
To test this idea, Koopman (1950) mixed two species of *Drosophila* that
mated almost as readily and successfully between the species as within the species. In each generation he discarded all hybrids, so that continuing generations of the mixture contained only progeny whose parents had not cross-
mated. In succeeding generations the number of hybrids decreased rapidly,
indicating a decrease in interspecific matings. According to these results there
was in the original experimental population of each species a mixture of individuals with a propensity to mate with their own species and those with a propensity to mate with the other. The fact that selection reduced the pattern of
mating so effectively indicates further that these propensities are behavioral
traits controlled by genetic factors. From this it is a reasonable hypothesis
that sexual isolation in nature arises from natural selection acting on sets of
multiple genes or heterozygous behavioral factors.

Soon after Koopman's experiments, Vaurie (1950, 1951) described a situation in the Eurasian nuthatches of the genus *Sitta* that fitted this concept
perfectly. *Sitta neumayer* is primarily European, *S. tephronota* Asiatic; the
two ranges overlap broadly (Fig. 11). Where these two sister species occur
separately, their beak size and head markings are virtually identical. In the
area of overlap, the species differ markedly in both beak size and head markings. It seems certain that where they overlap the two species have affected
each other in a manner that has accentuated their differences. The differences
in head color are presumably associated with sexual recognition and on that
basis can be explained by the application of Koopman's experiments. The differences in beak size are undoubtedly associated with differences in feeding
habits. Thus in *Sitta,* natural selection is producing a shift in character states,
termed *character displacement* by Brown and Wilson (1956).

A comparable situation was discovered by Blair (1955) concerning two
sister species of frogs, *Microhyla carolinensis* and *M. olivacea.* The ranges of
these two frogs, *Microhyla carolinensis* and *M. olivacea,* overlap along a narrow
band in Oklahoma and Texas. In this area of overlap the species hybridize in

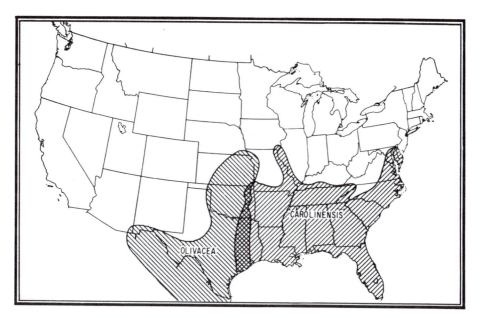

Fig. 5.12 Ranges of the two frogs *Microhyla olivacea* and *M. carolinensis*. (Adapted from Hecht and Matalas, and Blair, in H. H. Ross, *A Synthesis of Evolutionary Theory*, © 1962. Reprinted by permission of Prentice-Hall, Inc., Englewood Cliffs, N. J.)

at least some localities (Fig. 12). The chief factor which tends to reduce hybridizing seems to be differences in mating call. Through a study of mating calls, Blair found a most interesting phenomenon. In the area of overlap and the adjoining areas east and west of it, the calls of the two species differ greatly in the average of both frequency and duration and differ completely when these two factors are considered as a unit (Fig. 13). In distant nonoverlapping parts of the two ranges, however, the calls of the two species are almost identical.

A reconstruction of the past history of these two species indicates that *M. olivacea* was formerly more southwestern and *M. carolinensis* more southeastern, the two originally being allopatric. Following the last glacial dissipation, these two species spread northwestward and northeastward, respectively, eventually meeting in or near the present zone of overlap. Until the time of meeting the mating calls of the two species were presumably only slightly different, *M. carolinensis* as in E (Fig. 13), and *M. olivacea* as in G. Because of selection against hybrid individuals, the differences between the calls of the two species became intensified and now, in the overlap area, the call of *M. carolinensis* (A) is lower and shorter than that of *M. olivacea* (D).

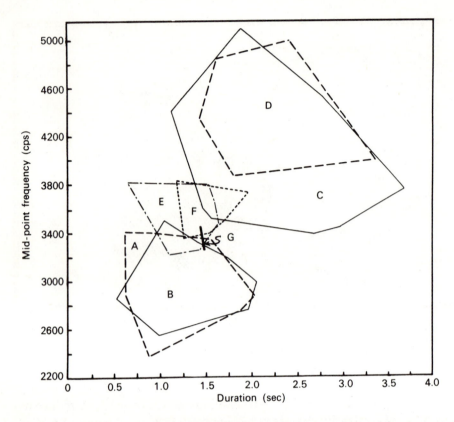

Fig. 5.13 Comparison of *Microhyla* mating calls by polygons constructed from scatter diagrams. A—*carolinensis* from the overlap zone; B—Texas-Oklahoma *carolinensis* from east of the overlap zone; C—Texas-Oklahoma *olivacea* from west of the overlap zone; D—*olivacea* from the overlap zone; E—*carolinensis* from Florida; F—presumed hybrids from the overlap zone; G—*olivacea* from Arizona. (From W. F. Blair, 1955.)

In Australia, Littlejohn (1959) found a comparable situation with seven Australian species of the frog genus *Crinia.* Of the seven species, four have become sympatric and in these the calls are sharply different. Calls of allopatric species exhibit only average, overlapping differences.

From these and more recent observations and experiments we arrive at an operational hypothesis of great importance: when allopatric sister populations evolve sufficient genetic difference that cross-matings will produce markedly lower numbers of viable progeny than intrapopulation matings, if the ranges of the two populations become overlapping, then a positive selection pressure will favor any genetic changes reducing or eliminating cross-mating. At this point in their evolution, the two populations will represent evolutionarily distinct lineages and will be separate species.

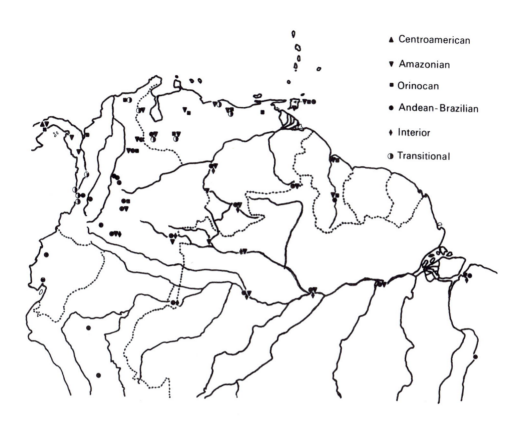

Fig. 5.14 Known geographic distribution of the semispecies of *Drosophila paulistorum.*
(From Spassky *et al.*, 1971.)

Because speciation by geographic isolation is a gradual process, one
would expect to find in nature species exhibiting transitional situations between any designated steps in the speciation process. Dobzhansky and
Spassky (1959) described an especially interesting situation involving the evolution of six incipient species within *Drosophila paulistorum,* a species rigidly isolated reproductively from its closest relatives. The incipient species,
later called *semispecies* by Dobzhansky and Pavlovsky (1971), are indistinguishable morphologically and unknowns can be identified only by crossing
with tester individuals from known colonies. Populations within a semispecies are completely interfertile, but laboratory crosses between semispecies
are either completely sterile or produce fertile females and sterile males. Two
of the semispecies have somewhat allopatric ranges (Fig. 14), but the ranges
of all of them overlap in northern South America and Panama (Spassky et al.,

1971). In many localities three semispecies occur sympatrically, different combinations of semispecies in various spots. In these localities the different semispecies are intersterile, whereas several of them from allopatric localities are partially interfertile. Although this evidence is admittedly meager, it suggests strongly that each semispecies is sufficiently incompatible genetically with any other, that when any two occur together, the two evolve sexual isolation with regard to each other. On this basis it would appear that the semispecies of *Drosophila paulistorum* represent evolutionarily independent lineages, in other words, distinct contemporaneous species. Dobzhansky and Spassky (1959) pointed out that it is possible through fortuitous crosses for genetic material of one semispecies to become incorporated into that of another semispecies, even though there is no evidence of this in nature. This phenomenon is in reality potential introgression. The totality of the *Drosophila paulistorum* example, however, indicates clearly that interpopulation crossing experiments may not settle conclusively the problem of "How many species are represented in a group of populations?"

Examples of genetically-tested plant groups emphasize the same point. Lewis (1953a, b) crossed members of geographically distant populations of *Clarkia deflexa* and found varying degrees of interfertility between them (Fig. 15). *C. deflexa* occurs in dense local colonies, has a low vagility, and a tendency to produce chromosomal rearrangements. These factors have produced a most puzzling array of populations in which neighboring populations appear to be evolutionarily distinct, whereas more distant populations are apparently genetically compatible. As Dobzhansky said about the semispecies of *Drosophila paulistorum*, at least some of these local populations of *Clarkia deflexa* are species in the state of being born. Some of these *Clarkia deflexa* populations would fit Dobzhansky and Pavlovsky's concept of "semispecies," and the definition of species in the sense of having reached the point of evolutionary independence. Other populations appear not to have reached such a point because they produce a fertile F_2 generation.

This *Clarkia* example emphasizes a difficult point in the recognition of species. Theorectically there is some point in the evolution of diverging populations at which they would be sufficiently different genetically that they would not recombine into a single interbreeding population even if intermingled. Techniques that will detect this point in instances when populations are near the borderline of passing from subspecies to species have not yet been perfected.

A Speciation Model

With the above background information a theoretical model of speciation by geographic isolation through range division can be devised. It represents a

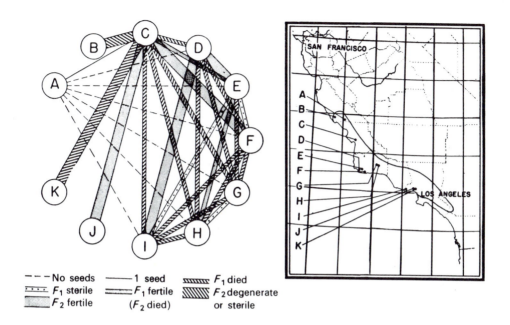

No seeds ——— 1 seed 〰〰〰 F_1 died
⋯⋯ F_1 sterile ═══ F_1 fertile ⧄⧄⧄ F_2 degenerate
▨▨ F_2 fertile (F_2 died) or sterile

Fig. 5.15 Fertility relationships of populations of *Clarkia deflexa* having the geographical distribution indicated. Progenies indicated as having died without issue were in some cases victims of disease; in others, they were not pollinated. (From H. Lewis, 1953*a*.)

process of gradual change which in nature could not be divided into sharply defined steps. Theoretically, however, in this gradual evolution of sister species from their original common parent, we can define certain points or way stations that are important concerning further developments in the evolving lineages. Minimally, five steps are involved (Fig. 16).

Step 1. The parent species with an essentially continuous range. If the range is small, the entire populations representing the species might be completely panmictic. If the range is large, genetic differences in different parts may be sufficiently great that distant populations may be intersterile. This phenomenon has been verified experimentally in several insect and mite species, and in certain birds and reptiles.

Step 2. The division of the range of the parental species into two or more parts sufficiently well separated that individuals of one part cannot reach other parts and mate with individuals in them. In short, populations in one part of the divided range are physically isolated from populations in other parts, with no or only sporadic gene flow between them.

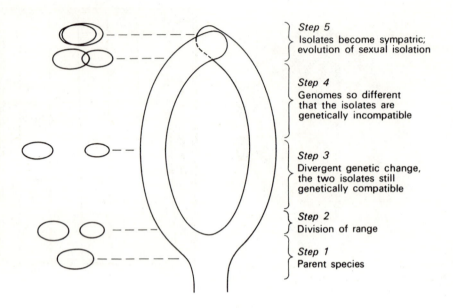

Step 5
Isolates become sympatric;
evolution of sexual isolation

Step 4
Genomes so different
that the isolates are
genetically incompatible

Step 3
Divergent genetic change,
the two isolates still
genetically compatible

Step 2
Division of range

Step 1
Parent species

Fig. 5.16 Steps in the evolution of sister species by geographic isolation and genetic change. Cross-sections are shown at the left.

The range division may occur as a result of any one event or a combination of several. Common ones include: submergence of a land bridge, isolating terrestrial populations on each end; emergence of a land bridge, isolating marine populations on each side of it; and changes in climate in areas having a varied topography.

Step 3. The gradual and divergent genetic change in the isolated units. Genetic change occurs continuously in populations of all species, and two factors ensure that it would normally be different in two geographically isolated ranges. First, it is certain that genetic changes in two isolated units would not all be the same, considering the chance occurrence of new mutant alleles and the many possibilities of gene recombinations. Second, the chances are great that the areas occupied by two isolated units would be different ecologically to at least some extent and that each area would therefore exert different selection pressures on the population unit in that area. Ecological differences might involve rainfall, humidity, temperature, and daily or annual combinations of these and other factors.

The net effect of divergent genetic changes and selection pressures would result in the isolated populations becoming progressively dissimilar genetically with the passage of time. In the early stages of change, the two populations would still be compatible genetically. During this step, if the ranges of the isolated units came together, the two populations would interbreed freely, produce viable, fertile progeny, and eventually form a single interbreeding population. This population might have more variability than the original parent population. At this point, the isolated populations would not be species, but might be designated as some sort of subspecific category depending on circumstances. This problem of names is discussed more fully in the next chapter.

Step 4. The evolution of genetic incompatibility between the isolated units. If genetic change in the isolated units progressed sufficiently, then when individuals of one unit were experimentally crossed with those of another, the hybrid progeny would be inviable or have reduced fertility. Dobzhansky (1951, 1970) listed many examples of these phenomena in animals. Stebbins (1950) did the same for plants. In early stages of genetic incompatibility, F_1 hybrids might be healthy and fertile, but the F_2 generation might be either sterile or have reduced viability. In more advanced stages, the F_1 hybrids might be healthy but partially sterile, then weak and wholly sterile. As genetic incompatibility increased still more, the hybrid individuals would die at various stages before maturity, in the embryonic stage, or at an early cleavage. Ultimately, the hybrid combination of gametes would not unite to form a zygote.

Hybrid genetic incompatibility is caused by a number of circumstances, discussed in detail by Dobzhansky (1970). Of great importance is the lack of meshing in a physiological sense of the genotypes of the two parents. Presumably the lack of meshing results in a breakdown of the biochemical pathways controlling development. In many crosses the chromosomes do not pair properly at meiosis, resulting in unbalanced gametes and zygotes that are usually inviable. In some species of *Drosophila*, the passage of foreign sperm in the vagina or sperm receptacle is slowed or completely obstructed, usually resulting in a lack of effective insemination (Patterson and Stone, 1952). Certain plants have an interspecific lethal gene that causes the early death of hybrid seedlings, but has no effect on unions within its own species (Hollingshead, 1930).

A most interesting and as yet poorly understood facet of hybrid sterility is the "infectious sterility" associated with symbionts or parasites found in *Drosophila* by Williamson and Ehrman (1967). Later Kernaghan and Ehrman (1970) found mycoplasma-like bodies in the testes of sterile males. Yen and Barr (1971) found that what had been described as cytoplasmic incompatibility in *Culex* mosquitoes (Lavan, 1959) was associated with infections of a rickettsia-like microorganism, possibly *Wolbachia pipiensis*, found in the maternal cytoplasm of the eggs and other tissues but not in the sperm. The *Drosoph-*

ila and *Culex* examples suggest that certain strains carry symbiotic microorganisms to which the strain is adjusted and which do not interfere with fertility. In hybrid combinations the symbiotic strain carried by the mother apparently has a disruptive effect on the male with resulting sterility.

If two geographically isolated sister populations had evolved to any of these stages of genetic incompatibility, and if their ranges came together, they would not fuse into a single interbreeding population, but would persist as separate breeding units; that is, they would be independently evolving lineages. This independence would result from one of two factors. First, the isolated units might by chance have evolved sexual isolation in addition to genetic incompatibility before the geographic congregation of the isolates, in which case there would already exist barriers to interbreeding when the sister species first came into contact. The longer the species were apart before contact, the greater would be the possibility of their possessing such barriers to interbreeding. Second, if no barriers to cross-mating had evolved when the species came into contact, the following step 5 would ensue.

Step 5. The evolution of sexual isolation. If two isolated populations evolve considerable genetic difference and if the ranges of the two become overlapping before sexual isolation between them has evolved, then two types of natural selection against hybrid progeny could come into play. First, if the hybrid progeny were genetically compatible but inefficient compared with the parent types, they would contribute less and less to succeeding generations. Second, if the two populations were genetically incompatible but yet did intermate freely, then any cross-mating individuals would contribute less to future generations than would their true-mating brothers and sisters.

Both of these circumstances, hybrid ecological inefficiency and genetic incompatibility, would result in gametic wastage which, in turn, would reinforce and accentuate any traits in either or both species that decreased interspecies mating. Such traits that decrease the chances of interspecies mating in sympatric species are termed *reinforcing mechanisms.* Stebbins (1950) and Dobzhansky (1951, 1970) gave a discussion and classification of these mechanisms in plants and animals, respectively. Both authors considered them to be a type of isolating mechanism and grouped them with the factors leading to genetic incompatibility which we treated above. Reinforcing mechanisms may precede the sympatry of previously allopatric species, but if not, they follow the sympatry of these species.

Reinforcing mechanisms may be grouped into several broad types:

1. *Ecological separation.* The two sister species become adapted to contrasting ecological types. For example, in the southeastern coastal plain, *Vernonia augustifolia* occurs on dry sand ridges whereas *V. gigantea* occurs on the

moist soil between the ridges (Jones, 1972). Presumably, the resulting micro-geographic isolation of the two species reduces cross-fertilization, which in turn would favor new genetic combinations tying each species more rigidly to its specialized niche.

Related plant species pollinated by different species of insects have frequently evolved different forms and colors of the inflorescence, presumably associated with reinforcing the attractiveness of the flowers to specific pollinators. The remarkable array of flower forms in certain orchids is an excellent example supported by a great deal of observation (Darwin, 1862; Ames, 1937). The refractive patterns for ultraviolet light shown by many flowers (Eisner et al., 1969) are undoubtedly evolutionary results of natural selection for various "targets" used as landmarks by pollinators.

Host specificity of prey species also functions as an ecological separation of populations. This phenomenon is especially notable in insects and parasitic animal phyla, and probably in fungi. Closely related species often have different hosts, or if they sometimes feed on the same host, they mate and reproduce on different hosts. Presumably if two species that were genetically incompatible but not sexually isolated occurred together on two or more hosts, there would be a selection pressure for any traits that led to the specialization of each species for a different host.

A common type of ecological separation involves temporal and seasonal factors. In these, the reproductive stages of the two species are active either at different times of day or at different seasons of the year. Examples are abundant in both plants and animals.

2. *Mechanical isolation.* Stebbins (1950) pointed out that in the genus *Asclepias,* the pollen sacs or pollinia are attached into pairs by a clip. When an insect visits an *Asclepias* flower, one of these clips catches on its legs and is carried to the next flower, where the pollinia fit into slits in the stigma. Each species differs in the size and shape of its pollinia and the corresponding slits of the stigma. It is difficult for the pollinia of one species to fit into the stigmatic slits of another, resulting in a great reduction in cross-fertilization.

Another type of mechanical isolation discussed by Stebbins concerns the growth of pollen tubes in the styles. In *Nicotiana* and some other genera pollen tubes of hybrids do not reach the ovaries, sometimes because the pollen tubes do not grow as well in styles of foreign species. This latter case would involve a chemical factor.

3. *Mating patterns.* In many groups of animals, related sympatric species have evolved specialized mating behavior, often with the evolution of distinctive color patterns and morphological traits especially in the male and the production of volatile sex attractants (pheromones) by one or both sexes. Part of

the pattern consists of male displays of visual cues, such as the peacock's strutting; various types of waggling and promenading, as in salamanders and fishes; song production, as in cicadas, crickets, and at least some fishes (Gerald, 1971); stroking or tapping of the female, as in some birds and insects; and other specialized types of behavior. Normally the female will not respond to or will actively reject a male of a foreign species.

Stebbins (1950) emphasized the circumstance that, in plants, sympatric species are isolated genetically by a combination of various reinforcing mechanisms acting in conjunction with varying degrees of genetic incompatibility. His discussion of the topic indicates that if genetic incompatibility has reached a high level, the reinforcing factors may be insignificant, but that if the genetic incompatibility is at a low level (for example, fertile F_1's, sterile F_2's), the reinforcing factors are usually highly evolved.

For this selective mechanism to work, it is necessary to suppose that at the time of species contact each species had a slightly different spread of genetic factors governing the trait even though the two spreads had a wide overlap. This would result in a different mean or mode of these determinants in the two genetic entities. Selection against gametic wastage would therefore remove the overlapping bands of gene frequencies, producing in each species a new distinctive mode for these characters. Figure 17 shows diagrammatically the probable steps in this process, discussed on p. 74. The speed would vary with the extent of cross-mating, the percentage viability of the hybrids, the differences in ecological or behavioral patterns at different stages of the process, and the rate at which new mutations are added to the genetic system controlling the characters involved in the reinforcing mechanisms.

The development of reinforcing mechanisms is thus seen to follow the same type of gradual evolutionary change that is characteristic of the other processes involved in the evolution of species by geographic isolation.

In summary, the processes leading to species fission by geographic isolation begin with ecological and geologic changes which bring about a division of the range of the parent species and the formation of allopatric species (species whose ranges do not overlap). These species may subsequently expand their ranges or reverse ecological changes may occur in the environment causing the ranges of the newly formed species to overlap. After this congregation of species, sexual isolation (if it has not done so before) will evolve as the final step in the evolution of mature daughter species.

Colonization

The idea basic to species division by colonization is that one or a few individuals of a species disperse across one of the peripheral areas unsuited to the species, reach a suitable ecological area as yet unpopulated by the species, and

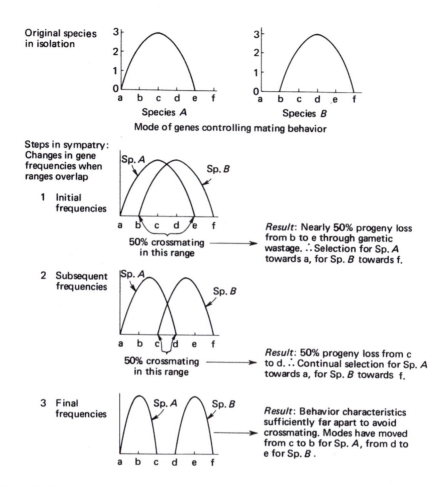

Fig. 5.17 Diagram of the theoretical steps in the evolution of sexual isolation between two species that are genetically incompatible and whose ranges overlap. Hypothetical units of behavior are indicated by the letters a through f. Note that when they were allopatric, species A had its mode at c and species B had its mode at d. (From H. H. Ross, *A Synthesis of Evolutionary Theory*, © 1962. Reprinted by permission of Prentice-Hall, Inc., Englewood Cliffs, N. J.)

colonize this latter area. The colony as established is cut off from all but infrequent gene exchange with its parent population. Consequently it develops in virtual isolation and evolves into a different species. Except for the mode of establishing the disjunct population, colonization presumably effects an increase in number of species in the same manner as does a division of the range.

The new colonial population would be different from a segment of a divided range in one significant respect. In a divided range, each segment would have a great deal of the genetic variability of the parental species. In a colonial population that arose from the successful establishment of one or a few individuals, the genetic variability would initially be only that of the colonists, and would presumably be much more restricted than that of the parent population. Compared with its parent, the colonial population would lack some genetic units completely, and the ratio of the other units would be shifted. Should selection pressures in the colonized area be different from those acting on the parent population, the colonial population would change genetically at a relatively rapid rate. This *founder effect* has been discussed by Mayr (1942, 1963), MacArthur and Wilson (1967), and others. The occurrence of speciation by colonization is proven by the distinctive species inhabiting oceanic islands which, although far from and never connected with the mainland, nevertheless support a varied biota that must have originated on the mainland. The biota of the Galapagos Islands, the Hawaiian Islands, and many others must have originated in this manner.

In addition to these striking cases of long-distance colonizations, short-distance examples are numerous. The list includes many West Indies insect species closely related to mainland species, the unusual insular subspecies of mice on the islands of British Columbia, and several birds and frogs on Tasmania. One might consider that colonial populations having an irregular but moderately frequent arrival of additional members from the parent population would not evolve into distinctive species. The continuous arrival of new immigrants might be considered a sort of link which would preserve enough genetic continuity between colony and parent range to nullify the differential effects of partial isolation. Yet the fact remains that new species have evolved or subspecies have differentiated on islands a comparatively short distance from continents.

The natural attainment of genetic isolation by insular species under these conditions is demonstrated by the phenomenon called double colonization. One example concerns the hornbill bird species *Acanthiza pusilla* and *A. ewingi*. The latter is an endemic Tasmanian species which evidently arose from an early colonization by the Australian *A. pusilla*. Tasmania can now be reached from Australia by two successive island-to-island jumps of only 50 miles each. During each Pleistocene glacial period, these 50-mile intervals were mostly or completely above water, hence these dispersals of *Acanthiza* may represent successive range disjunctions rather than colonizations.

A. pusilla occurs both on the Australian mainland and Tasmania, but on Tasmania it does not interbreed with *A. ewingi* (Mayr, 1942). Mayr concluded that colonists of *A. pusilla* dispersed to and colonized in Tasmania and there evolved into a distinct species *A. ewingi* which is now genetically isolated from its original Australian parent. J. A. Moore (1954) reported a similar case of insular speciation involving two frogs *Crinia signifera* and *C. tasmanica*. The former occurs both on the Australian mainland and Tasmania, and individuals from one area cross successfully with mates from the other area. *C. tasmanica* occurs only on Tasmania. Although a close relative and an obvious derivative of *C. signifera*, *C. tasmanica* will not cross with it and apparently evolved from an old colonization of Tasmania by *C. signifera*. Mayr (1942) gives additional examples of double invasion in many parts of the world.

The use of the word "double" might in itself imply that in a "double colonization" only two dispersals from continent to island did indeed occur. The short distance from Australia to Tasmania makes this circumstance highly unlikely, nor is it a necessary postulate to explain the case. It might be argued that if other invasions had occurred, either more than two species would be present on the island, or the colonists interbred freely with the island population. This latter possibility leads to a more plausible explanation of insular species evolution based on the premise of probable multiple colonizations.

Let us assume a hypothetical example involving a bird such as the hornbill of Australia which might occasionally reach an island such as Tasmania either by being swept there by storm or by fortuitous wandering. Successive colonizations could follow these theoretical steps:

1. With animals, the initial colonization will begin when a male and female or an inseminated female reach the island and raise a family. In monoecious, self-fertilized plants, a single fertilized propagule can effect a successful colonization; in dioecious plants a propagule producing each sex is necessary. In both plants and animals in which there is an alternation of both sexual and apomict generations in the life history, a single apomict vagrant can produce a successful colonization.

2. The genetic composition of the island colony will change independently of the parent colony because the ecological setting will be slightly different, thus exerting slightly different selection pressures. The original genetic make-up of the colony will be different from the parent population because it will contain only the genetic make-up of two parent individuals rather than the entire gene pool of the parent population. Later, genetic drift will be different in the parent and colonial populations. Thus a genetic divergence will arise between the parent population and the colony. Vagrants reaching the island during the initial stages of this divergence would presumably interbreed freely with members of the colony. The total number of the colonists would be so large in proportion, however, that genetic differences of an occasional vagrant would be absorbed and diluted with rapidity.

3. The genetic divergence between the colony and the parent population would reach a point at which only limited genetic compatibility existed be-

tween them. Vagrants reaching the colony would mate with members of the colony, but the offspring from these unions would be chiefly sterile; hence few of them would be able to reproduce normally. Thus if the number of vagrants remained constant, their effect on the gene pool of the colony would diminish with the passage of time.

4. When the genetic divergence reached such proportions that the colony was completely incompatible genetically with the parent species, vagrants would undoubtedly continue to mate with members of the colony, even though all progeny of such matings would be completely sterile. As previously cited observations in *Drosophila* and *Microhyla* indicate, it is commonplace for allopatric species to cross-mate when members of closely related species accidentally meet, and further it appears that *sexual* or behavorial isolation (at least between closely related species) is a result of contact between the main populations of the species involved. Because our island colony, now a new species, would be in contact with its parent species only through an occasional vagrant, there would be only negligible selection pressure toward the evolution of sexual isolating mechanisms. Also, because vagrants would presumably arrive only one at a time in any one locality, they would mate with a local member of the island species rather than have an opportunity to mate with each other. Even if an odd family of vagrants became established, their progeny would be rendered ineffective by cross-mating with members of the island species. At this stage the parent species would become reestablished as a second invasion on the island only if a small band reached it and if the band kept together as a breeding unit.

5. When the divergence between the island and parent species became sufficiently great that a state of sexual isolation prevailed between them, the situation would revert to Step 1, when a pair of the parent species could start a continuing second colonization of the island.

There would be a period embracing all of Steps 1 through 4 when sporadic vagrants could have reached the island without leaving any tangible trace. If a "double invasion" were ultimately accomplished, the record would give evidence only of the first and last arrivals.

The series of steps listed above are patterned after the requirements of animals such as birds or termites in which both the male and female must be present during the breeding season. The situation would differ in certain respects in the case of animals such as other insects, snakes, or mammals in which a vagrant gravid female could establish a family. With these, the more important factor would be population size. A large island population would tend to swamp the relatively minute proportion of vagrant progeny either by amalgamation into the population or, if hybrid progeny were sterile, by exterminating them because of chance cross-mating.

Theoretically this series of colonizing events could be repeated again and again, ultimately giving rise to many colonial species. Wilson (1959) postulated that such multiple invasions have been an important factor in the evolution of the ant fauna of Melanesia.

The most remarkable example of speciation by colonization within a group of islands was the evolution of the fly family Drosophilidae in Hawaii. As now known, the Hawaiian drosophilid fauna comprises almost 500 species, of which 96.4% are endemic. Phylogenetic studies made to date indicate that the colonies probably originated from only two and perhaps only one colonizer, almost certainly from the Asiatic region. Apparently successive colonizations between the eight main islands of the archipelago have led to successive phylogenetic branchings and the evolution of additional distinctive species. It is estimated that an additional 200 endemic species still await discovery on these islands (Hardy *in* Carson *et al.*, 1970).

Of this endemic fauna of roughly 700 species, each is restricted to a single island, except that Maui, Molokai, and Lanai must be treated as one biological island because they have been connected at least twice during the Pleistocene due to lowering of the sea level. The 700 endemic species would have required at least that number of successive interisland dispersals sufficiently long ago that the established colony would have evolved into a species different from its parents. It is probable that the actual number of successful colonizations was far in excess of this minimum 700. According to the model of geographic speciation by colonization outlined above, a moderately steady number of infrequent colonization did go undetected during the course of the speciation process, until the sister species on the various islands had evolved sexual isolation in addition to genetic incompatibility.

Possibilities of new species production by colonization. The possibilities for increase in number of species by colonization are manifold, and could be considered as a product of the multiplication of dispersal possibilities by the number of barriers to species range extension. The manner by which plants and animals become dispersed across inimical ecological areas is extremely varied and has been discussed at length by Allee et al. (1949), Allee and Schmidt (1956), and Darlington (1957). By far the greatest amount of dispersal which might lead to colonization is passive: by rafting following floods, by wind, or by other agents. Some animals may cross barriers actively. Flying birds may wander or be blown over mountains or water barriers, or roam far from their normal range. Fish occasionally may swim through waters which they do not normally occupy, and mammals may similarly wander long distances into new

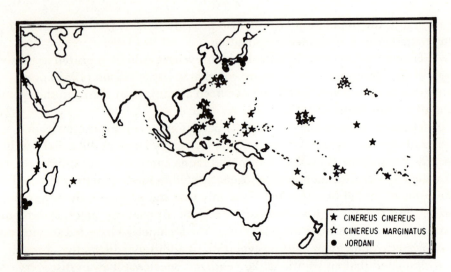

Fig. 5.18 Known distribution of two species of conger eels of the same genus *Conger*. Note the disjunct African populations of both species. (From R. H. Kanazawa, 1958)

territory. The present disjunct distribution of certain conger eels probably came about by such wanderings (Fig. 18).

The number of barriers is extremely difficult to assess. In general every distinctive ecological area of a continent or an ocean is a barrier to those species which cannot exist in it. However, attempts to correlate patterns of species production and present ranges with existing barriers encounter difficulties imposed by a dynamic past.

As Stebbins (1950) pointed out, colonizations which today seem improbable because of the great distances involved may have occurred at some time in the past when the barriers between areas were narrower than at present. Also, areas now disconnected may have been connected in the past. On the basis of present geography the Santa Cruz jay *Aphelocoma insularis* appears to be a colonist. It is possible, however, that formerly Santa Cruz Island was connected to the California mainland and that the Santa Cruz jay is the product of a past range disjunction of the mainland species *A. californica* race *obscura* (Pitelka, 1951).

Instances of geographic disjunction therefore fall into three categories: range disjunction, colonization, and a great number of doubtful cases which might be the product of either process.

Time and the Speciation Process

In assessing the status of daughter entities involving speciation by geographical isolation, it would be most convenient if we had a measure of time whereby we could say "Lineages separated for so many years are species, lineages separated by less are not yet species." The consensus of evidence is that we cannot make such a statement. The length of time required for isolated sister populations to develop evolutionary independence in a genetic sense appears to vary greatly between different taxonomic groups and even between closely related species. The moss *Grimmia patens* (Steere, 1938) and the amphibian *Rana sylvatica* (Smith, 1957) suggest disjunctions of at least 10,000 years without discernible change. The peculiar range of the sword moss *Bryoxiphium norvegicum* (Fig. 8) suggests an even longer disjunction, perhaps of several million years, with little discernible change (Steere, 1937; Löve and Löve, 1953). Because of ease of spore transport, the moss examples may represent more recent dispersals. Judged on the basis of morphological similarity, at least a 35,000 year separation has not produced distinctive species in many boreal North American stoneflies and caddisflies each of which has a northeastern population and a widely disjunct Ozarkian population. A stonefly example is *Allocapnia pygmaea* (Fig. 19).

Certain disjunct Ozarkian populations, however, have evolved into species which are distinctive on a morphological basis. Three examples are the winter stonefly *Allocapnia sandersoni* and the caddisflies *Hydropsyche piatrix* and *Glyphopsyche missouri.* Each of these species is most closely related to a sister species occurring considerably to the north and east (Ricker, 1952; Leonard and Leonard, 1949). In all three examples the Ozark species differs from its northeastern relative in slight but constant morphological differences comparable in both type and magnitude with differences between closely related pairs of allied sympatric species.

In all three examples, however, it is at present impossible to ascertain the time that the populations giving rise to the Ozark endemic species dispersed to and became isolated in that area. It could have been prior to the last glaciation or prior to any previous glaciation. We don't know. Uncertainties in ascribing particular biological events to particular geological events (explained more fully in Chapter 8) make it very difficult to equate evolution of distinctive sister species with geologic time.

Dobzhansky and Pavlovsky (1966, 1971) demonstrated that partial hybrid sterility can occur in less than five years. In 1958 a Llanos strain of the Orinocan semispecies of *Drosophila paulistorum* produced fertile males and females when crossed with other populations of the semispecies. By 1964 the Llanos strain had changed so that when crossed with the same populations all the males produced were completely sterile. The cause of the infertility was not known, but injection experiments suggested the possibility that it might be caused by the Llanos strain having acquired a new mutant type of symbiont.

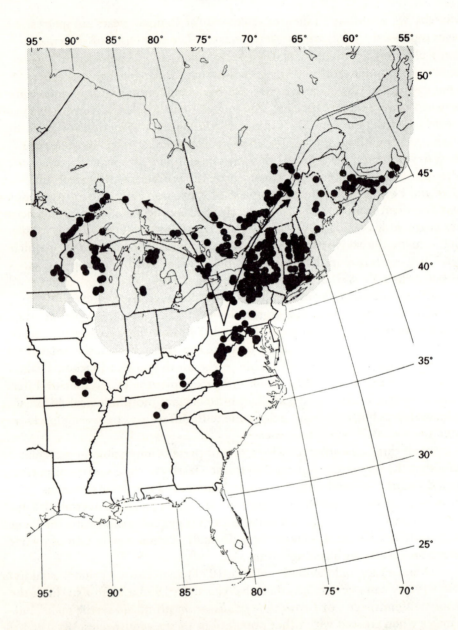

Fig. 5.19 Distribution of *Allocapnia pygmaea*. Arrows indicate probable routes of post-Pleistocene dispersal. The southwestern populations in Missouri occur only in artesian rivers, and together represent a glacial relict. (From H. H. Ross and W. E. Ricker, "The classification, evolution, and dispersal of the winter stonefly genus *Allocapnia*," *Ill. Biol. Monog.*, 45:1-166, 1971. Copyright © 1971 by the University of Illinois Press and reprinted with permission.)

At times various maxims have been proposed concerning relative rates of evolution; for example, that the rate should be faster for organisms with a short generation time than for those with a long generation time. Simpson (1944) pointed out that among mammals opossums had a short generation time, elephants a much longer one, yet the rate of evolution in elephants has been many times more rapid than that in the opossums. He also drew attention to the circumstance that size, longevity, and other features are correlated phenomena, and that no general function between any of these and evolutionary rate could be discerned.

These same phenomena have been cited as affecting the rates of speciation by geographic isolation. For example, in eastern North America at least seven species of the leafhopper genus *Erythroneura* live only on sycamore trees, the genus *Platanus*. All seven leafhopper species form a single monophyletic unit that almost certainly evolved on sycamore from an original sycamore-inhabiting parent species. During this period the sycamore species of eastern North America evolved only little and did not speciate. Because at present all seven *Erythroneura* species occur over almost the entire range of the native sycamore, it would seem that the sycamore populations were isolated in the past to the same extent as were the parental leafhopper populations. It appears, however, that the presumably isolated sycamore populations never progressed beyond Step 3 of the preceding speciation model, whereas the leafhoppers evolved to Steps 4 and 5. This and numerous similar examples indicate that in general insects form new species by geographic isolation more rapidly than do plants.

On the other hand, it could be argued that length of generation time was the important factor (1/3 year for *Erythroneura,* about 10 years for sycamore); or population density (up to many thousands to one for *Erythroneura* vs. sycamore), which is a space function of individual size. But the opossum-elephant example contradicts both the size and the generation time correlation. Perhaps the only reasonable conclusion is that rate of speciation is different in various groups; this rate is a value to be discovered in each without prejudice of prior assumptions.

Chromosomes vs. genes. One facet concerning rate of speciation seems to be well documented, the role of chromosomal versus genic changes. The best evidence about this comes from plants.

Although relatively slow rates of evolving genetic incompatibility appears to be the usual circumstance in plants, there are notable exceptions. In the genera *Holocarpha* (Clausen, 1951) and *Clarkia* (Lewis and Roberts, 1956), closely related species and even local populations within the same morphological species are rigidly intersterile. Cytogenetic investigations disclosed that

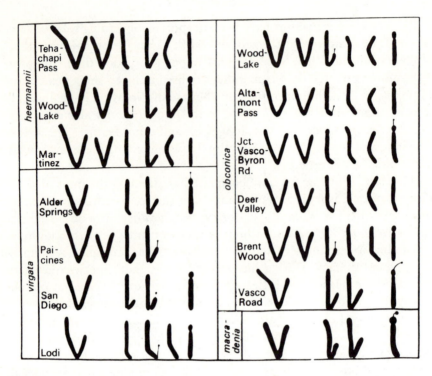

Fig. 5.20 Chromosome morphology of four species of the plant genus *Holocarpha* illustrating one chromosome of each pair. Note the differences between local populations of the same species. (Reprinted from Jens Clausen: *Stages in the Evolution of Plant Species.* Copyright © 1951 by Cornell University. Used by permission of Cornell University Press.)

each entity had a different chromosomal configuration (Fig. 20), and it was undoubtedly these differences that accounted for the sterility of hybrid individuals. Both authors concluded that evolution of genetic incompatibility is slow when it depends only on the accumulation of genic changes but may be extremely rapid when chromosomal rearrangements occur.

Wallace (1968) proposed the following speciation model that could lead to unusually rapid speciation with a minimum of geographic isolation. Theoretically, border populations of a species represent gene pools that have reached the limit of their fitness in the face of increasing ecological vicissitudes, and are prevented from building up genotypes able to disperse into the adjacent ecologically inimical areas because the border populations are continually being swamped genetically by genotypes from the heart of the range that are less fit for border existence (Carson, 1955; for further discussion, see Chapter 12). Wallace suggested that due to weather fluctuations border populations could become isolated from the main range, evolve genotypes better fitted to the border-and-beyond area, and in the process also acquire additional chromosomal rearrangements that resulted in genetic incompatibility between the isolated border population and the parent population. A reverse climatic fluctuation would bring the two populations into contact, with a resultant selection through gametic wastage for noncrossability between individuals of the two populations. During successive weather fluctuations, the isolated population, now free from swamping effects of the parent population, would rapidly diverge from it genetically in response to the challenges of a new environment, and any recontact with the parent population would reinforce the sexual isolation between them. Under these circumstances the isolated border population could evolve into a new species in a minimum amount of time. Determining the exact amount of time required would nonetheless be difficult.

In many respects Wallace's model is similar to the evolution of *Clarkia lingulata* from a border colony of *C. biloba* (Lewis and Roberts, 1956). This model, however, invoked a massive destruction of hybrid genotypes due to unusually xeric seasons, a process that Lewis (1962) called speciation by *catastrophic evolution*.

Cryptic Species

A special circumstance that is often puzzling to the beginner concerns species that are genetically and sexually distinct but which cannot be differentiated on the basis of observed characteristics. For example, the freshwater copepods commonly referred to as *Cyclops vernalis* include several species reproductively isolated from each other but not identifiable on the basis of known diagnostic characteristics. For identification, individuals of several species must be crossed with individuals from known sets of test stocks. Several examples of cryptic species have been found in *Drosophila* (Dobzhansky and Spassky, 1959), in other animals, and in plants (Anderson and Whitaker, 1934).

In most discussions of this topic (e.g., Selander, 1969), the terms *sibling* and *cryptic* are erroneously considered synonymous. *Sibling species* are sister species arising from the same ancestral species; whether or not they can be diagnosed by conventional systematic methods is of no concern. *Cryptic species* are those that cannot be identified by conventional means; it is not necessary that they be sibling species. The term *conventional means* is an arbitrary one meaning exactly what it says. For example, the early examples of cryptic species in *Drosophila* were identified originally on the grounds that crosses between members of populations seemingly identical morphologically would not produce offspring. Later, morphological differences were discovered between these species, after which they were no longer cryptic, although they proved to be sibling. The same circumstance is true of certain morphologically similar crickets first identified by song differences, then discovered to be reproductively isolated. At this stage these species were considered cryptic. When audiovisual techniques were discovered by which the songs could be registered visually with accuracy, the species were no longer "cryptic." Their differences could be seen by "conventional methods."

Speciation by Hybridization

Simple or isoploid hybridization is here defined as hybridization between species of the same ploidy (isoploids), that is, between two $2n$ species or between two $4n$ species, and so on, without the subsequent occurrence of polyploidy. The obvious hybridization between many isoploid distinctive plant species (many previous papers summarized in Anderson, 1949; Stebbins, 1950, 1959) early gave rise to the hypothesis that hybrids between different species would produce populations having the potential to evolve into distinctive new species separate from both parents.

Thanks primarily to Anderson (1949), it is now acknowledged that interspecific hybridization between distinct lineages leads not to the evolution of additional species but to changes in the two lineages that hybridize. In this process, termed *introgressive hybridization* by Anderson, one or both hybridizing parents, through backcrossing with hybrid individuals, may add to their genic complement certain genetic constituents of the other (see p. 60).

A few instances have been cited as examples of speciation by simple hybridization. The first example involved three Balkan firs of the genus *Abies*. The common fir of central Europe, *Abies alba*, extends through the western Balkans to northern Greece, and *A. cephalonica* extends through the mountains of central and southern Greece. Between these two species and extending considerably eastward occurs a highly variable entity *Abies borisii-regis* which is in general intermediate between *A. alba* and *A. cephalonica* and which is considered to be a hybrid species with *A. alba* and *A. cephalonica* as its parents (Mattfeld, 1930; Stebbins, 1950).

A second example involves three species of the American silkworm moths of the genus *Platysamia* (Sweadner, 1937). *P. euryalis* occurs along the west coastal region, *P. gloveri* along the more eastern Rocky Mountain chain. The third species, *P. kasloensis,* occurs between the northern extensions of the other two species and is in most respects intermediate between them. *P. kasloensis* appears to be a hybrid entity between *P. euryalis* and *gloveri* (Fig. 21). It has, however, gone further in becoming more evolutionarily distinct than has *Abies borisii-regis*. *P. kasloensis* is only partially interfertile with its supposed two parents. No sexual isolation exists, all forms intermating freely. Few of the F_1 generation reach maturity and those that do are low in vigor. Hybrid females are sterile, but hybrid males will backcross with either parent,

Fig. 5.21 Distribution of the species comprising the silkworm genus *Platysamia*. The species *P. kasloensis* is considered to be of hybrid origin, having arisen from *P. euryalis* and *P. gloveri.* (From W. R. Sweadner, 1937).

hence some interspecific genetic interchange is possible. In addition, *P. kasloensis* has evolved some distinctive though small characteristics of its own.

In contrasting the *Abies borisii-regis* case with the *Platysamia kasloensis* case, there is a strong suggestion that the latter species, after it originated as a hybrid population, was isolated geographically from both its parents long enough to evolve considerable genetic dissimilarity and its own distinctive characteristics.

In the *Drosophila virilis* complex the cytogenetics of the subspecies *D. americana americana* can be explained only as the result of hybridization between *D. americana texana* and *D. novamexicana* (Patterson and Stone, 1952). The hybrid subspecies *americana* occupies such a large range with so little overlap with the subspecies *texana* that it can justifiably be considered an incipient species resembling the *Platysamia* example.

More recent experimentation with plants throws a new light on speciation by simple hybridization. Grant (1966) was able to produce a new species of the genus *Gilia,* reproductively isolated from either parent, by careful selection of the progeny of a few individuals that were produced by an almost sterile F_1 hybrid.

Grant, however, considers that the evolution of a new and genetically isolated species by simple hybridization requires so many fortuitous factors for success that it probably occurs only infrequently in nature. The same conclusion is apparently true of hybrid species only partially isolated genetically, such as *Platysamia kasloensis.*

Hybrid Species in Plants and Animals

The idea of hybrid species in plants has been well founded and acknowledged by botanists for decades. Perhaps this viewpoint was accentuated by the relative ease of crossbreeding plants (as compared with animals) and growing the various hybrid combinations. The concept of hybrid species is largely ignored by most zoologists. As a result, little effort has been expended in the investigation of this phenomenon in animals. Because of what seems to be mosaic inheritance in certain insects, the hybrid nature of some animal species was first demonstrated with a high probability in this group of animals. Later work included vertebrates.

The possibilities of hybrid species as a vehicle of unusual and rapid evolution in animals was stressed by Ross and Ricker (1971) with regard to the origin of the anomalous winter stonefly *Allocapnia minima* in eastern North America, reminiscent of the predictions of Edgar Anderson. More recently, Meyer (1970) has emphasized the omni-directional evolutionary possibilities of hybrids by producing flowers with carpelloid (very primitive) petals in an interspecific hybrid of a cotton species. The consequences of this hybrid are unnerving from a logical point of view: Where hybrid species are involved, evolution can go backwards. From a practical standpoint, we have as yet no evidence that these reversed-evolution mutants give rise to new and distinctive species. But this does not mean that they could not. Rather it means that should we suspect evidence of evolutionary reversals, we should investigate these phenomena with unusually detailed scrutiny.

Speciation by Ecological Isolation

There are many instances in which two or more sister species are sympatric, having nearly identical ranges, but in which each species occupies a different ecological situation in their essentially common geographic range. In two sympatric African species of *Anopheles, A. melas* occurs only in brackish water, *A. gambiae* in fresh or nearly fresh water. In three extremely closely related sympatric species of North American flycatchers of the genus *Empidonax, E. trailli* nests in swamps and thickets, *E. minimus* in open woods, and *E. flaviventris* in conifers. In the coastal plain of the southeastern United States, *Vernonia angustifolia* occurs on dry sand ridges whereas *V. gigantea* grows on the moist soil between the ridges (Jones, 1972). In the plant genus *Salvia,* several closely related sympatric species flower at different times. In many sympatric sister species of insects, one species feeds only on one host while its relatives feed on different ones.

From these and innumerable other known examples of the ecological isolation of related species occurring within the same range there arose the suggestion that ecological isolation was a mechanism producing new species. It was argued that individuals in habitat *A* would tend to breed more with each other than with individuals in habitat *B,* and that this differential interbreeding would set the stage for genetic divergence and eventual new species formation between the populations occupying the two ecological extremes.

In most instances there is little reason to believe this would happen (Mayr, 1947). If one ecological extreme was located some distance from its opposite with an unpopulated intervening area, speciation would be by geographic isolation. If the ecological extremes interfingered geographically or were ends of a sliding scale, such as earlier and later seasonal activities, there would seem to be no way to prevent the formation of heterozygous populations that would have no sharp break genetically corresponding to ecological gradients.

Disruptive Selection

Several models have been proposed in which genetic differences associated with ecological differences would be under selection pressures favoring the extremes, or homozygotes, over the heterozygotes. The result postulated for this disruptive selection is the production of new sympatric species that would be ecologically distinct.

In a series of experiments with *Drosophila melanogaster* allowing random mating, Thoday and Gibson (1962) were able to produce disruptive isolation resulting in two sexually isolated strains by selection against heterozygotes between two lines, one having low numbers, the other high numbers, of sternopleural bristles. These results indicate that sympatric isolation is possible under certain required circumstances:

1. The existence of two different ecological situations A and B with little or no area intermediate between them.

2. A heterozygous species in which one homozygote was better fitted to situation A than the heterozygote or the other homozygote, and the other homozygote was better fitted to situation B than the other two genetic combinations.

3. A slight genetic difference in mating behavior associated with fitness for each ecological situation.

Wallace (1968) pointed out that (1) and (2) would be necessary to bring about an elimination of hybrids, and (3) would be a prerequisite for establishing sexual isolation. He summarized many unsuccessful attempts to repeat Thoday and Gibson's results, and concluded that the probability of sympatric speciation in nature by this model was remote.

Using house flies as a model, Pimental et al. (1967) developed a model of sympatric speciation based on the apparent evolution of subpopulations ovipositing differentially in two offered foods, banana and fish meal. Disruptive selection was achieved by removing from each subpopulation those eggs that were deposited in the "wrong" food. In about a year each of the two selected subpopulations had evolved a marked preference for its own food. Hybrids lacked this preference, indicating a genetic basis for the food selection of each strain. Further tests of this model are in progress (Pimental, *in litt.*).

Discussing pollination systems as isolating mechanisms in flowering plants, Grant (1949) presented a model for sympatric speciation in angiosperms based on the species-specificity of many bees. According to this model, an obvious mutant affecting flower color or shape might become established in a population. If the new type of flower happened to be selected and visited by a bee species other than those visiting the parent type, homogamy would result; in other words, the most similar individuals within the population would preferentially be mated to each other. Grant also mentioned that hybridization might also lead to sympatric speciation.

Basing his ideas on the plant genus *Penstemon*, Straw (1955, 1956) combined the idea of homogamy and hybridization to arrive at a most plausible model of sympatric speciation. He observed that garden-produced hybrids of the two most different species he had studied resemble another distinct species growing in the same area. All three species have different pollinators. He postulated that hybrids occasionally produced would be visited by neither pollinators of the parents; if another pollinator found them attractive, homogamy would ensue and there would be set in motion selection pressures that would eventually result in the evolution of a distinctive, sympatric hybrid species.

Ecological Breaks

Two kinds of ecological isolation can afford sharp ecological breaks and offer highly probable mechanisms for the formation of new sympatric species, *temporal isolation* and *host isolation*.

Temporal isolation. The type of isolation considered here is one that would divide the population of a species so that, in the same area, part of it would breed only at one time of year and part of it only at a different time, and that furthermore the progeny of each part would continue their parents seasonal timing. Under such circumstances, the two populations would be genetically isolated and would presumably evolve into different species.

In a large number of species, the production of sexual forms occurs continuously for weeks or months. If the active sexual form were short-lived, as in many flowering plants and insects, those produced earliest would not be able to mate with those produced latest, and vice versa. The earliest and latest individuals would therefore be isolated temporally. But because of the continuous production of individuals, there would be genetic panmixia between individuals over the entire time span in the same fashion that panmixia occurs geographically over a large area.

Two types of life history can provide absolute time disjunctions between different sympatric populations of the same species:

1. Annual cycles, in which a short appearance of sexual forms may be shifted one or two seasons.

An excellent example of an early stage in this process is afforded by the California annual tarweed *Madia elegans*, which at low elevations has both a spring- and a fall-flowering race (Clausen, 1951). The spring form develops no rosette but flowers quickly in March to May and withers after its seeds are ripe. The fall-flowering form requires the entire spring to develop a dense basal rosette of leaves, and does not flower until August, two months after the spring form has dried up. There is no possibility of direct gene exchange between these spring- and autumn-flowering populations.

A similar phenomenon has apparently happened in at least two genera of insects, resulting in the evolution of new species. In the sawfly genus *Neodiprion*, whose adults normally appear in spring, two lineages arose having fall-emerging adults. In the cricket genus *Achaeta,* which normally overwinters as eggs, one lineage arose in which the late nymphs are the overwintering stage. The result of these changes in part of a parental population would result in two annual adult emergences separated by an entire season.

2. Multi-year cycles.

Certain insects have a life history spanning several years, yet some adults appear each year. Many species of large scarab beetles, for example, have a two- or three-year life span, but a brood emerges each year. There is a possibility that each year class is effectively isolated from the others temporally and genetically, and that each of these classes may evolve into a separate species. The potential temporal isolation has been well established for the most famous multi-year-cycle insect — the periodical cicada, the genus *Magicicada* of eastern North America. There are two types with 13- and 17-year life cycles, respectively, and each has numerous year-class emergences or broods. Although each type has a total contiguous range, each brood differs slightly to considerably in range and greatly in numerical abundance. Because of widespread interest in these conspicuous and noisy insects, enough information is available to demonstrate remarkable temporal isolation between broods. The evolutionary implications as they relate to speciation are being studied by Drs. T. E. Moore and R. D. Alexander.

Host isolation. The prevalence of host-specific organisms in diverse goups of plants and animals has long stirred interest in the idea that in host-specific groups new species could arise by some sort of host-transfer mechanism. Host specificity is pronounced in bacteria, fungi, viruses, higher plant parasites such as mistletoes, mites, and literally hundreds of families of insects — fleas, lice, parasitic wasps and flies, leaf-feeding Hymenoptera, beetles, Homoptera, and many others. Those who would detract attention from host-transfer speciation have generally held to the idea that all new species arise by geographic isolation and that competition between species is the cause of host specificity.

The first highly probable case of new species evolution by host transfer was provided by Rothschild and Clay (1952). They established beyond reasonable doubt that a population of rodent-infesting fleas had become established in the nests of burrowing owls that were predators of the rodents, and that these owl-infesting populations had evolved into a new distinctive owl-specific lineage or species. It is of especial significance that in order for this event to have occurred, both the owls and their prey rodents *had to be sympatric.* Many examples implying this host-transfer mechanism have been enumerated (Emerson, 1949). The most extensive have been those involving the Nearctic species of the vagile leafhopper genus *Erythroneura* (Ross, 1962), in which there is evidence of some 150 host transfers in a fauna of about 400 species.

Studying fruit flies of the genus *Rhagoletis,* Bush (1969) postulated that several Nearctic pairs of sibling species arose sympatrically through a host-transfer mechanism. Of unusual interest is his analysis of a new host race of *Rhago-*

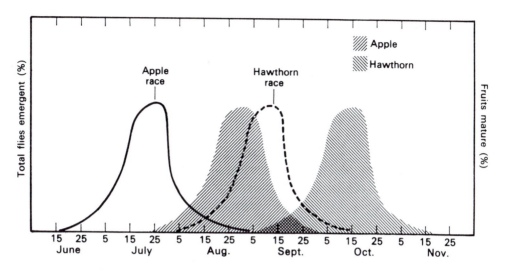

Fig. 5.22 Adult emergence period of the apple and hawthorn races of the fruit fly *Rhagoletis pomonella*. The cross hatched areas represent the approximate period of fruit maturation when larvae are leaving the fruits and pupating in the soil. (From G. L. Bush, 1969).

letis pomonella in eastern North America which became established on introduced apples a little over 100 years ago from the original *Crataegus*-infesting form. Since that time, the apple race has evolved an adult emergence period whose mode is six weeks earlier than the *Crataegus* race, presumably in response to selection pressures associated with earlier host fruit maturation. The result is that the emergence periods of the two races have only a slight overlap (Fig. 22). How much, if any, gene flow occurs between them is not known. The two races also have slight but significant differences in relative body size, number of postorbital bristles, and ovipositor length.

Bush discussed five factors that could reduce gene flow between two newly-established sympatric host races:

1. If mating occurs only on the host, then selection of the host by different individuals would automatically result in assortative mating.

2. If the two hosts had temporal differences in the times at which the structure used for food was available, this would provide an allochronic separation that would reduce the possible period of inter-race matings.

3. Disruptive selection according to the models of Levene (1953) and Smith (1966) acting on genes associated with habitat selection, pleiotropism, assortative mating (as in the first factor), and modifier genes could lead

to a drastic reduction or complete elimination of gene flow between extreme genotypes, without the influence of geographic isolation.

4. If the new host race becomes conditioned to, and it and its progeny become accustomed to, a particular host stimulus, this would (in conjunction with factor 1 above) reduce inter-race mating to a minimum. Bush listed several instances in which host conditioning has been observed in insects.

5. The "magnitude" of the host shift would have an important effect on the rapidity of divergent evolution in the new race. If the host shift was to a species in the same genus whose biochemistry was presumably similar to that of the original host, there would be little selection in the new race for a change in genetic makeup. If the shift was to a phylogenetically distant species, perhaps to one in a different family, the new host would presumably have a much different biochemistry. The result in this situation would be selection pressures favoring changes in genetic makeup, thus enhancing the operation of all the preceding four factors.

Bush concluded that the operation of all five factors can lead rapidly to the formation of new host races and species. Although the examples given include only insects, the principles involved should apply also to other relatively host-specific groups, such as fungi, parasitic worms, and various parasitic Arthropoda such as Acarina and certain Crustacea. Certainly, even a cursory examination of the host relationships of these groups suggests that speciation by host transfer is a widespread phenomenon.

Speciation by Polyploidy

Up to this point the discussion has concerned speciation by the slow change of entire populations through genic change or chromosomal rearrangements. We now turn to a genetic mechanism called polyploidy by which new species come into existence almost instantly.

Small changes in chromosome number occur commonly in many groups of organisms. Extra chromosomes are usually duplicates which have resulted from unequal migration at meiosis. Lower chromosome numbers in viable species, on the other hand, usually result from the fusion of chromosomes. The new number often becomes typical of an entire species, and a set of closely related species may have gametic chromosome numbers of 5, 7, 8, and 9, respectively.

Some related species differ in multiples of a basic set of chromosomes. Certain violets in the genus *Viola* have gametic chromosome numbers of 6, 12, 18, 24, and 36, respectively. Whole new sets of the original 6 chromosomes have become added successively to the genome of the ancestral species. This addition of whole sets of a chromosome complement is the phenomenon of *polyploidy.*

Polyploidy arises as a mutation in a single cell. It occurs during cell division when either the chromosomes divide, but the cell does not or when all the daughter chromosomes migrate into one of the dividing cells. The result is one cell containing twice as many chromosomes as the parent cell before cell division began. Dividing cells artificially treated with the drug colchicine (which inhibits *cell* division, but not *chromosome* division) often produce large number of polyploid cells. When polyploidy occurs during reduction division, it results in unreduced gametes.

The number of chromosomes of an organism is expressed as either the gametic number, for example, $n = 15$, or the zygotic number, for example, $2n = 30$. If one sex has an additional chromosome, the $2n$ number will not be a simple doubling of the n (which is really $1n$) number. When polyploidy occurs, the extent of polyploidy is expressed as multiples of the basic n number. Thus if a plant in which the zygotic number $(2n)$ is 30 gives rise to a polyploid individual in which the zygotic number is 60, this would be four times the gametic number 15, and would be noted as $4n$, which is a tetraploid. The gametes of this $4n$ individual would be $2n$, which in this case would be 30.

If only one gamete in a population undergoes a polyploid mutation (that is, it becomes $2n$ instead of $1n$) it may mate with a normal $1n$ gamete and form a triploid zygote $(3n)$. The zygote will usually grow normally, but will be sterile or nearly so. At meiosis, the odd number of chromosomes interferes with the successful production of viable gametes. As a result, single polyploid mutations do not result in new species of bisexual organisms.

If, however, a somatic polyploid mutation occurs in a cell that gives rise to a flowering branch, all the gametes of that branch will be polyploid. If the polyploid pollen from one flower fertilizes the polyploid ovules of another on the same branch, the zygote will be $4n$, or tetraploid. This zygote will normally be balanced genetically and will usually produce viable gametes, all with the meiotic reduction chromosome complement of $2n$. These $2n$ gametes will produce viable offspring if they mate with each other, but will produce only sterile triploids if they mate with $1n$ gametes of the pre-polyploid parent. The new tetraploid population is therefore intrafertile, but reproductively isolated from its closest relatives. In every evolutionary sense, it is a new species.

The first reported origin of a polyploid species occurred at Kew Gardens, England (Digby, 1912). A perennial hybrid of a cross between *Primula verticillata* and *P. floribunda* was viable, but sterile. After some time a polyploid shoot arose that produced fertile polyploid flowers.

Since that time, thousands of bisexual polyploid species have been discovered in plants (Stebbins, 1950), but relatively few in animals. The earwigs or Dermaptera have some tetraploid and hexaploid species (Bauer, 1947). In a few hermaphroditic animal groups, chiefly flatworms, leeches, and earthworms, the chromosome number of certain groups of related forms almost certainly indicates polyploid species. Beçak et al. (1966) reported the first vertebrate examples, two species of South American frogs. Recently Wasserman (1970) has demonstrated without reasonable doubt that the North American tree frog *Hyla versicolor* is polyploid. The reason for this drastic difference between plants and animals is that most higher plants have two one-mutation mechanisms for producing polyploid species whereas animals have none.

The two plant mechanisms are:

1. a somatic polyploid mutation of a monoecious species in a cell producing any part of the plant which bears gametes of both sexes. If the mutation occurred in the zygote, the entire plant would be polyploid; if in a branch or shoot, all flowers of the shoot would be polyploid. If the plant was self-fertile, any self-pollinated seeds of plant or shoot would be polyploid and produce polyploid progeny.

2. a somatic polyploid mutation in a cell giving rise to all or part of a spore-bearing frond. All the spores in the affected part would be polyploid. Considering the millions of spores produced by a large frond, this would give a fantastic rain of polyploid spores, each with the potential of becoming a polyploid gametophyte thallus. Even if these were dioecious, enough polyploid individuals should be close enough together to produce some polyploid zygotes and establish a new polyploid species.

In each mechanism, polyploid gametes of both sexes are produced at the same time and in close proximity to each other.

In hermaphroditic animals such as earthworms, a polyploid mutation of the zygote would theoretically be such a one-mutation mechanism, although in these animals the two sexes of gametes are normally produced at different times. The relatively high incidence of polyploid species in the hermaphroditic groups suggests that self-fertilization does occur, even if only rarely.

In most metazoan animals and dioecious spermatophyte plants, there is no known one-mutation mechanism for producing polyploid species. In these organisms, polyploidy would seem to require two polyploid mutations, one resulting in male and the other in female gametes. The simplest situation would be two zygotic mutations occurring in the same batch of eggs. Even then the chances of those two mutant individuals happening to mate together would be remote. Perhaps some peculiar environmental condition might cause many eggs in a batch to undergo polyploid mutation. White (1954) suggested that a species could become parthenogenetic, then polyploid, then regain its bisexuality. In groups such as aphids with series of parthenogenetic generations between the sexual generations, it would theoretically be possible for a parthenogenetic female to become polyploid and eventually give rise to polyploid males and females in the same locality and at the same time. But no polyploid species have so far been recorded in this group of insects.

Another reason often proposed for the paucity of animal polyploids is the greater metabolic complexity and lesser development flexibility of animals compared with plants.

Two entirely different types of polyploids occur, *autopolyploids* in which the two chromosome sets come from the same species, and *allopolyploids* in which each set comes from a different species. In some botanical literature, these types are termed *autoploids* and *amphiploids,* respectively. The latter situation arises when the fruiting structures of an interspecific hybrid become polyploid, as in the example of *Primula* in Kew Gardens. In nature relatively few species are autopolyploids. In her fascinating studies on fern cytology, Manton (1950) found a small number, but a far greater number of the fern polyploids were of hybrid origin.

Allopolyploidy in plants has unusual evolutionary significance in two respects. Because of it, whole sets of polyploid species have evolved to form a series of polyploid levels. A simple example in which the species differ only slightly is the European triad of fern species in the genus *Polypodium* (Manton, 1950). In these forms three fertile levels of ploidy occur: diploid, tetraploid, and hexaploid, with gametic numbers of 37, 74, and 111, respectively. Sterile perennial triploids and pentaploids ($3n$ and $5n$) also occur in nature where the parent species overlap. More complex examples are found in the section *Chamaemelanium* of the genus *Viola* in which four levels of polyploids occur above the diploid level (Clausen, 1951). Several levels occur also in the genus *Clarkia* (Fig. 23), but with irregular chromosome numbers because of the different numbers in the original parent species.

There is an apparent limit to the extent of polyploidy compatible with viability and fertility, probably associated with increasing problems of physiological ontogenetic adjustment that accompany an increase in the number of genes affecting the same processes. The great bulk of plant polyploids are tetraploids or hexaploids, but occasionally higher levels occur. In the higher plants, natural species seldom exceed 12-ploids (Stebbins, 1950; Clausen, 1951). In mosses, possible decaploids are the highest natural level so far found (Steere, 1954); in ferns, the highest reported gametic chromosome numbers of 250 to 260 may represent as high as a 20-ploid (Manton, 1950).

Polyploidy offers unusual possibilities for the combination in one lineage of characteristics possessed by two or more lines so different that they would not produce fertile hybrids. Chiefly involved in this role would be intergroup hybrids and intergeneric hybrids. An example of the former is the grass *Bromus carinatus* complex which represents an allopolyploid between the sections *Bromopsis* and *Ceratochloa* of *Bromus.* An example of the intergeneric type

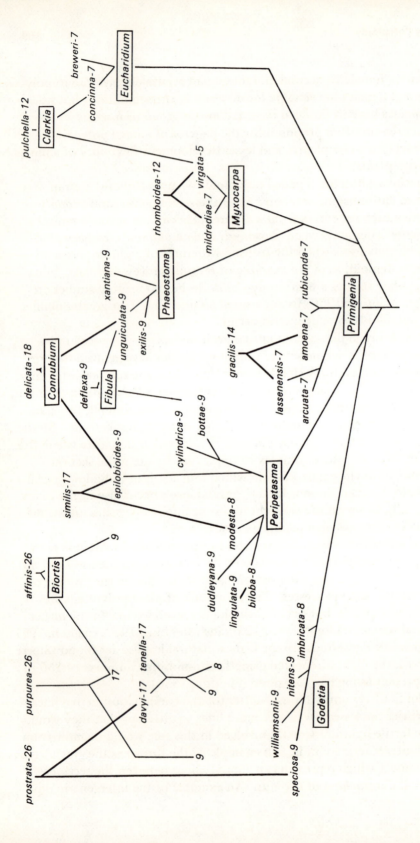

Fig. 5.23 Family tree of the plant genus *Clarkia*, showing the large number of polyploid species arising from interspecific hybrids. (From H. Lewis and M. E. Lewis, 1956.)

is domestic wheat, an allopolyploid of the hybrid between *Triticum dicoccoides* and *Aegilops speltoides*. Although naturally occurring polyploids of proven intergeneric parentage are few, many divergent types probably arose by this method.

Polyploidy has been of tremendous significance in increasing the number of plant species. The lowest possible estimate of polyploid species is 25 percent with the possibility of this figure reaching 35 or 40 percent. The percentage differs widely in various plant groups. In the mosses (Steere, 1954, 1958) and angiosperms (Stebbins, 1950) some families have little or no polyploidy, whereas others have many examples; only in the ferns (Manton, 1950) are the bulk of the families almost entirely polyploid.

The many thousands of polyploid species do not *each* represent a new phylogenetic line that arose by the process of polyploidy. Thus in *Clarkia* (Fig. 23) eight of the ten polyploid species each represent a separate polyploid origin, but the two species *C. davyi* and *C. tenella* both arose from a single polyploid line which divided into the two sister species by isolation and nonpolyploid genetic divergence. In the horsetail genus *Equisetum*, the 11 fertile species listed by Manton (1950) all have a gametic chromosome number of 108, surely polyploid, but all 11 species have undoubtedly evolved by ordinary nonpolyploid processes from a common polyploid ancestor having a gametic chromosome number of 108.

Naming Polyploid Lineages

Because of the universally reduced fertility of triploids and other odd-n individuals, it is obvious that every even-n polyploid in a series is distinct in an evolutionary sense from other members of the series. Each even-n polyploid ($4n$, $8n$, $12n$, etc.) therefore represents a distinctive lineage and in an evolutionary sense is a separate species. In many instances in which the polyploids exhibit diagnostic characters other than these cytological ones, such as size of pollen grain, length or shape of leaf, etc., systematists have recognized them as such and given a name to the taxa involved. All allopolyploids fit this situation, as also do a small number of autopolyploids.

However in many autopolyploid series the individuals of different ploidal levels are otherwise remarkably similar. Here there is hot debate, so far restricted to botanists because no comparable instances are yet known in animals. The great majority (e.g., Raven et al., 1968; H. Lewis, 1969) argue against giving specific names to these cytologically distinct, but morphologically similar, entities. A few (e.g. W. Lewis, 1969) point out that what we are really seeking is a deeper scientific understanding of the biota, and not simply an easy way out for the taxonomist who is dealing primarily with dead specimens. While W. Lewis shows some caution in full acknowledgment of all polyploids as separate species, he does propose that those cryptic cytotypes of evolutionary significance be recognized as species. If we are to make systematics evolutionary, we should recognize as species *all* polyploids in a series. The situation is exactly comparable to the *Drosophila* species that were once considered sibling or cryptic, and for years could be identified only on the basis of crossing with pedigreed stocks. The same is still true of species in the *Cyclops vernalis* group of copepods (Price, 1958).

Giving distinctive epithets to each member of a polyploid series may assist in advancing nonsystematic investigations. An example concerns present studies of the pollination by bees of the creosotebush, *Larrea divaricata*, in southwestern North America. Yang (1970) found that this plant comprised three parapatric chromosomal types now called races: the diploid ($n=13$) primarily in the Chihuahuan desert, the tetraploid ($n=26$) in the Sonoran desert, and the hexaploid ($n=39$) in the Mojave desert. Each race has distinctive morphological and physiological traits and no triploids or pentaploids has so far been found. Linsley (1972) as a spokesman for the bee teams feels that, from an evolutionary, ecological, and morphological standpoint, these are distinct species. In studies of comparative bee sociology regarding these three species, it would be helpful to have legal nomenclatural names by which to record data and conclusions.

Giving scientific names to cryptic polyploid entities that are successful evolutionary units would draw attention to their existence. This in turn should lead to advances in our knowledge of the role of these entities in ecological communities.

APOMICTIC LINEAGES

In both plants and animals, species arise in which there is no sexual fertilization between different individuals of the population. The result is a lack of genetic mixing, a phenomenon called *apomixis*. The progeny of succeeding generations form radiating pedigrees or *clones* quite different from the interwoven mesh of cross-matings found in bisexually reproducing organisms (Fig. 4b). Except in cases of hybrid apomicts, there is much less opportunity for diverse genetic recombinations than in bisexual species.

As has been pointed out by numerous authors, bisexually reproducing species form the long, continuing lineages from which the present diversity of life has arisen. In these, there is a periodic alternation of reduction division which results in cells having a gametic chromosome number and also, at some point of the life history in the production of male and female gametes. The process of fertilization, the union of male and female gametes, results in a zygote or fertilized egg having essentially double the gametic chromosome number. Apomixis is a process by which a female parent gives rise *without normal fertilization* to progeny of the same chromosome number. In other words, dual sexuality is circumvented in one way or another. The number of mechanisms associated with circumventing the normal sexual cycle in apomicts as a whole, and the intricate web of relationships within various complexes of apomictic species is almost incredible. A great deal of interesting and illuminating material on these topics is given by Gustaffsson (1946-47), Stebbins (1950), Fryxell (1957), and Grant (1971) for plants and White (1954) for animals, with especial reference to cytogenetic, genetic, and evolutionary aspects. The following account stresses material of especial importance to systematists.

Mechanisms of Apomixis

Apomicts can be classified into two principal types, one involving only vegetative reproduction, the other involving the production of an embryo.

Vegetative reproduction. The simplest type occurs in unicellular organisms such as the trypanosomes and the *Fungi Imperfecti* in which reproduction is always by simple fission without the occurrence of sex cells at any point in the life history. In a more advanced type, known in the annelid worm *Enchytraeus fragmentosus,* the body fragments periodically and each fragment develops into a full-grown worm; no sexual stages have been found in this species (Bell, 1959). Many plant species reproduce only by stolons, runners, or bulb offsets. A large proportion of these plants are polyploids that produce flowers which appear to be normal but which are actually sterile. Thus certain triploid lilies are flowering polyploids that reproduce only by offsets from their bulbs. Instances of fragmentation and vegetative reproduction undoubtedly evolved while the parent species was also reproducing sexually. Later in its evolution, the species lost its sexual behavior.

Embryonic reproduction. Three distinct mechanisms fall under this category. In the first mechanism, a diploid somatic cell of inner tissues of the ovule divides and forms an embryo. Only mitotic divisions occur, resulting in diploid embryos. The known examples occur in a wide variety of plants, including *Citrus* and many ferns. No animal examples are recorded.

In the second mechanism, a 2n germ cell gives rise to a 2n egg or embryo by circumventing meiosis. This mechanism, called *agamospermy* in botany and *parthenogenesis* in zoology, is the one found in the great majority of both animal and plant apomicts. It is advantageous competitively in relation to besexual species because all the individual's reproductive resources go into ovarian formation and in dioecious organisms all the individuals produce female offspring. The mechanism allows the reproduction of otherwise infertile hybrid, polyploid, and aneuploid genotypes, including those with high chromosome numbers in which meiosis and doubling would fail.

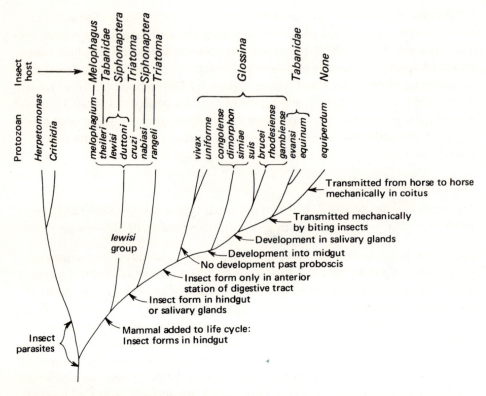

Fig. 5.24 Probable phylogeny of certain trypanosomes; an example of extensive evolutionary development within an apomictic group. The large right branch is the genus *Trypanosoma*. (From H. H. Ross, *A Synthesis of Evolutionary Theory*, © 1962. Reprinted by permission of Prentice-Hall, Inc., Englewood Cliffs, N. J.)

A truly remarkable array of mechanisms have evolved by which normal fertilization is suppressed. In some cases, meiosis never occurs in gametogenesis; in other cases, the products of meiosis reunite before the reduced cells divide, thus restoring the 2n chromosome number; in others, the early cleavage nuclei unite in pairs to bring about the same result; in still others, the egg unites with a polar body to restore the chromosomal complement. These and other ways of circumventing normal fertilization are described by Stebbins (1950) and White (1954).

The third mechanism involves the loss of the sexual generation in a life history that normally includes a succession of parthenogenetic generations. This occurs in aphids of the genus *Adelges* and in the tropical populations of some species in other genera. When this happens, the life cycle is a repetitious series of agamospermic or parthenogenetic forms. Many of the *Fungi Imperfecti* undoubtedly belong to this category.

Splitting of Already Apomictic Species

In apomicts having only mitosis in their life histories, theoretically the occurrence of a single viable mutation would result in a new distinctive type. In apomicts having meiosis plus chromosome number restitution, there might be a possibility for gene recombinations; hence the elimination of heterozygotes by natural selection might be necessary for the formation of a distinctive species. Although the splitting of apomictic species seems so simple, in general only a few cases occur here and there among the various taxonomic groups. In many instances either only one species in the genus is obligatorily apomictic, such as the sawfly *Diprion polytomum,* or a genus contains only a single species which is apomictic, such as the sawfly genus *Endelomyia.* The four known apomictic (parthenogenetic) species of black flies (Simuliidae) occur in three genera, indicating that in these insects the phenomenon occurred independently three and perhaps four times, with no observed subsequent increase in the number of species in any one of the phylogenetic lines (Basrur and Rothfels, 1959). In these instances it is highly probable that no species splitting has occurred since the evolution of the original apomictic parent.

In other instances of apomixis subsequent splitting has occurred. One of the most extensive apomictic evolutionary developments in animals is found in the trypanosomes. These protozoans evidently arose from a form parasitizing insects, in which a mammal host became included in the life cycle. By successive stages (Fig. 24), the protozoans evolved into a form in which their entire life history occurs in the mammalian host (Hoare, 1957). Some 20 species and many other races and strains are known, ranging from the primitive *T. lewisi* group to the highly specialized venereal species of horses, *T. equiperdum.* Observations made to the present indicate that this entire development is the result of strictly apomictic evolution, because no sexual stages have ever been observed in this branch of the Protozoa. Quite possibly similar series occur in simple plants, especially the Fungi Imperfecti, but little seems to be known about the relationships of these forms.

Hybrid and Polyploid Apomicts

In both plant and animal groups having a fairly large number of apomict species, the phenomenon seems to be associated with hybridization combined with polyploidy.

Certain parthenogenetic species of fishes have been associated with a hybrid origin (Hubbs & Hubbs, 1932), although the ploidy levels have not yet been determined. Certain genera of salamanders and lizards have many parthenogenetic species and in most instances these are diploids or triploids of hybrid origin (Uzzell and Goldblatt, 1967; Uzzell, 1969; Lowe and Wright, 1966; Lowe et al., 1970a; Maslin, 1968, 1971. Lowe et al. (1970b) have adduced ev-

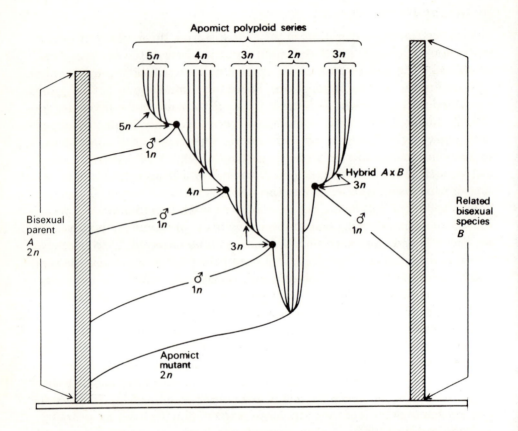

Fig. 5.25 Diagram of the probable origin of apomict polyploid species in certain weevil groups through occasional fertilization of apomict ova by male gametes of bisexual species.

idence that the hybridization pattern leading to the parthenogenetic species of the lizard genus *Cnemidophorus* has been highly complex, involving hybridization between species belonging to different species groups within the genus. In at least the amphibian and fish apomicts, another unusual feature occurs. The ova of the apomict species will not develop unless the mother is mated by a male of a parental or related species. No fertilization seems to occur, but the male sperm apparently transmits some type of stimulus to the parthenogenetic ovum. This mechanism is termed *pseudogamy*. In this sense, the apomict species are sexual parasites of the parental species.

More complex examples occur in the weevils. In the weevil subfamilies Otiorhynchinae and Brachyderinae numerous apomictic polyploid species have evolved, including triploids, tetraploids, and pentaploids. It is probable that in these weevils (Fig. 25) a diploid parthenogenetic species evolved first and that polyploidy followed as the result of occasional fertilizations of the parthenogenetic eggs by sperms from a male of one of the related diploid bisexual species (White, 1954).

The dozen or more species or races of the American genus *Graphognathus,* the white-fringed beetle, are all parthenogenetic (Buchanan, 1939; Warner, 1971). The number of characters shared in common suggests that these species form a monophyletic unit in which there has been an increase in number of species since the progenitor became parthenogenetic. Unfortunately, no cytogenetic evidence concerning the evolution within *Graphognathus* is yet known.

The most complex examples occur in the plants. In certain genera, large numbers of hybrid and polyploid apomictic species have evolved which together form morphological and ecological "bridges" between the parental diploid sexual parents (Stebbins, 1950). Such a situation has been investigated in detail for many genera including *Rubus, Poa, Crepis, Hieracium* and *Taraxacum* (Babcock and Stebbins, 1938; Stebbins, 1950).

Naming Apomictic Species

As with sequential species, apomict species can be named in only a subjective fashion. The chief criterion is whether or not a band of clones is sufficiently different from others that it would be profitable to give it a name from the standpoint of recording data about it. Certainly an obligatory apomict should be recognized as distinct from its bisexual progenitor. Stebbins (1950) has emphasized that apomicts may be of great value in understanding the evolution and dispersal routes of taxa under study. But some situations are so complex (see Stebbins, 1950, pp. 409–411) that a system of names other than the conventional may be needed, possibly using a lettering and numbering system.

It is a matter of record, however, that many apomicts have biological, ecological, and distributional characteristics that make their recognition and identification both illuminating and simple. The best example I have encountered is the trypanosomes (Fig. 24).

SPECIAL CASES

Two situations differ from the pattern of species dealt with above.

Infertile Perennial Hybrids

In her classic work on fern systematics, Manton (1950) reports an ongoing population of unusual composition. It comprises infertile hybrids of two species occurring together over a wide range. Although infertile, these hybrids are perfectly viable vegetatively and each individual lives for many years. Each year additional hybrids of the same species are added to the population. As a result, this sterile population is a definite ecological entity and competitor of other species in the community. Because it cannot reproduce itself, it cannot be considered a phylogenetic line. But from the standpoint of community relationships this hybrid population is a continuing viable entity. The practice of the botanists in giving such hybrids a distinctive name is an aid in accumulating and organizing data on this interesting facet.

An example of a statistical type of population composed of infertile perennial hybrids is afforded by the mule. This infertile hybrid of the horse and the donkey is maintained by selection by man because it combines the strength of the horse and the climatic adaptability to hot weather possessed by the donkey. Although not now as numerous as formerly, in the days before the common use of tractors and trucks, the mule was much more abundant than either of its parents, especially in agricultural districts of the United States.

Bacteria

In his discussion of microbial systematics, Mandel (1969) points out that in this group there has as yet been adduced no evidence that they form discrete phylogenetic lineages comparable to those found in eukaryotic organisms. Rather than denying such a phylogenetic structure, Mandel proposes tests by which such a structure might be detected and demonstrated. His opinion is that such a structure does exist. The very fact that most bacteriologists agree as to the identity of specimens in relation to taxa of long standing certainly supports this view. It will be interesting to see what future investigations indicate.

6

The Interpretation of Contemporaneous Species

On the basis of species analyzed in Chapter 5, it is possible to enumerate criteria concerning the recognition of species as they represent cross-sections of phylogenetic lines (lineages that are evolutionarily distinct).

First, a sharp distinction must be made between sequential and contemporaneous species. Sequential species can be associated only after the available contemporaneous species for each time level have been determined. The association of sequential species is therefore a special study area of phylogeny, treated in Chapters 7 and 8.

Contemporaneous species are those occurring at the same time level, whatever it is. We incline to think of contemporaneous as today, but it applies to any time level. The species found intermingled as fossils in a uniform stratum of any geologic age were contemporaneous and all knowledge of them indicates that they can be analyzed systematically in the same manner used for analyzing the present-day biota.

Sequential species are steps in the evolution of a single phylogenetic line. Contemporaneous species indicate the number of phylogenetic lines or lineages occurring at any particular time. The remainder of this chapter deals with contemporaneous species, which are the result of phylogenetic branching.

INITIAL STUDY PROCEDURES

The problem involved in the determination of contemporaneous species boils down to this question: given a number of collections from various localities, how many species do we have? How many independently evolving lineages do these specimens represent?

The first procedure is to orient and characterize the material assembled for study. This assumes that the investigator has become interested for any one of various reasons in a certain group of organisms and has made or accumulated study material over as wide an area as possible, prefereably the entire range of the group. Three initial steps are needed:

1. Check the study material through appropriate identification aids and establish that the study unit, comprising collections from various localities, is a closely related cluster of individuals representing either species, subspecies, or populations.

2. Become acquainted with the character differences of the group already mentioned in the literature, and initiate an intensive search for previously unnoticed character differences. Today there is an ever-expanding array of new tools for this pupose, chief among them being advances in electron microscopy and in microchemical analyses. Establish the amount of difference of each character within the study taxon. Set up a scale of conditions or states for each character having different states.

3. Sort individuals as to different character states, and record the results as a tabulation expressing the distribution of all the character states.

Two types of results are shown in Tables 1 and 2. In Table 1, the character states have a random distribution. Only specimens 2 and 8 are alike, and there is no complete correlation between any two character states. Thus while all *clefts* are *smooth*, two other *smooths* are *serrate*. A random distribution of this type suggests a single interbreeding population.

In Table 2 a great deal of order is present. Specimens 1, 4, 6, and 7 differ from specimens 2, 3, 5, 8, and 9 in leaf shape, flower color, stem pubescense, and seed shape; both sets of specimens exhibit both states for seed character. The concordance in the four character states is an excellent indication that the set including specimens 1, 4, 6, 7 and that including specimens 2, 3 5, 8, 9 represent two species, both heterozygous for seed texture. If tabulations of larger samples taken from various parts of the range show the same correlations, then the two sets would be classed at least tentatively as two species.

The more concordant the specimens are with respect to maintaining the differences in character states found between the two sets, the greater would be this possibility. Additional characters to investigate would include other morphological traits, as well as biochemical, behavioral, and cytogenetic ones. Frequently the character states between sets differ by only minute degrees and are detected only by careful comparison.

The amount of order in the characteristics of the study specimens may vary in different parts of the range, with various ecological conditions, or at different seasons. Also, intermediates between certain character states may occur in certain specimens or entire collections. These intermediates should be recorded and any associations with geographic or other external factors noted.

Differences noted between sets may be due to different genotypes, but some might be environmental variation expressed as ecophenotypes. The

Specimen No.	Leaf	Flower	Stem	Seed texture	Seed shape
1	serrate	blue	hairy	smooth	elongate
2	serrate	blue	spiny	rough	elongate
3	serrate	white	hairy	rough	round
4	serrate	white	spiny	smooth	round
5	cleft	blue	spiny	smooth	elongate
6	cleft	white	hairy	smooth	elongate
7	serrate	white	hairy	rough	elongate
8	serrate	blue	spiny	rough	elongate
9	cleft	white	spiny	smooth	elongate

Table 6.1 Random distribution of character states in nine specimens. Character states were selected by flipping a coin.

same could be true of some differences characterizing the two sets. It is therefore important at this stage to be as certain as possible that the character states used to designate sets of specimens represent genetic rather than ecological variation (see p. 28).

In attempting to determine the species status of the delineated sets of specimens, the investigator encounters one of two circumstances. In one, the sets are represented by living material amenable to various types of experiments, including especially hybridization tests or growing in uniform cultures. For these sets the species status may be inferred by direct reference to the evolutionary speciation processes outlined in Chapter 5. In the other circumstance, the sets are represented by only preserved material or material not amenable to hybridization tests. For these latter sets the species status can be inferred only indirectly from the processes outlined in Chapter 5.

Specimen No.	Leaf	Flower	Stem	Seed texture	Seed shape
1	serrate	blue	hairy	smooth	elongate
2	cleft	white	smooth	rough	round
3	cleft	white	smooth	smooth	round
4	serrate	blue	hairy	rough	elongate
5	cleft	white	smooth	rough	round
6	serrate	blue	hairy	smooth	elongate
7	serrate	blue	hairy	rough	elongate
8	cleft	white	smooth	rough	round
9	cleft	white	smooth	smooth	round

Table 6.2 Nonrandom distribution of character states in nine specimens. Character states were selected to simulate an orderly distribution.

SPECIES ANALYSIS WITH EXPERIMENTAL MATERIAL

After the sets have been delineated, reproductivity tests are made to discover whether each set is bisexual or not. This is done by determining the functional sexual stages present in each set. A variety of techniques may be used. In plants, seeds (if present) can be subjected to germination tests, the presence of sexual stages noted in fungi, and various cytological methods used to determine the presence of meiosis. In dioecious organisms, the absence of the male sex and the production of progeny by unmated animals or unpollinated plants can be observed. The following results are possible when comparing two sets or categories.

1. *No functional sexes in either category.* Both categories would be apomict clones reproducing asexually, by fragmentation, stolons, bulb offsets, conidiospores, or in some other manner. In plants, some asexual species have apparently normal flowering parts, but the seeds are shriveled and inviable. Whether or not to designate these clones as species would be a subjective decision of the investigator.

2. *Only functional females present in one category, no sexual stages in the other.* Both of these are apomicts, the category having no functional sexes being an asexual species, the other a unisexual species (called *agamospermic* in botany, *parthenogenetic* in zoology; for further explanation, see p. 64). If the two are very similar with respect to character states, it is likely that the asexual species evolved from the agamospermic or parthenogenetic species.

3. *Only functional females present in both categories.* Both apomicts, these categories would represent one or two unisexual entities. In types of unisexual reproduction having meiosis, there is a possibility of allelic segregation and different character states in the same line; hence it would be necessary to observe the progeny produced by individuals of the two categories. If the character differences bred true in each category, then two distinctive strains or species would be present.

4. *Only females functional in one category, both sexes functional in the other.* Under these circumstances, two strains or species would be involved: one unisexual; the other, bisexual.

5. *Both sexes functional in both categories.* Each category would be potentially at least one bisexual species. Further tests to try to determine the number of species follow the discussion of apomict species.

APOMICT SPECIES

In the above examples only two categories were illustrated for the sake of simplicity, but in nature large numbers of closely related apomict species may be encountered, as in the plant genus *Crataegus,* the protozoan trypanosomes, and certain weevil genera. After these apomict species have been sorted to units by character states, it is desirable to investigate their cytogenetics, suggested by the large number of known polyploid apomicts in plants, in a number of insects, and in a few salamanders (see p. 115).

When living material of apomicts is not present, the only recourse is to sort the specimens into sets according to character states, then to give each set a designation. Depending on the author, they may be termed species, races, biotypes, or other designations.

It is often difficult to be sure that a species is unisexual on the basis of dead or unreared material. In flowering plants, abortive pollen in herbarium specimens would give a possible indication of problems in meiosis, or of unisexuality. In insects especially, males may occur for a very short period compared with females and be missed completely by collectors. In many instances the association of the two sexes remains undiscovered for decades; eventually when the sexes are associated, the males may prove to be abundant. A special case involves the Fungi Imperfecti; periodically someone discovers the sexual stage of one of them and it is removed from the list of asexual species.

BISEXUAL SPECIES

Sets of bisexual organisms suspected of being different species are tested in three ways:

1. cytogenetic examination;   2. inter-set hybridization;

3. intra-set hybridization.

Cytogenetic Examination

If the gametic chromosome numbers of the sets exhibit a progression in multiples of a single number (e.g., n=6, n=12, n=24), and if the chromosomes of the higher numbers sort into a corresponding number of morphologically similar groups, then the different sets are a polyploid series. The species could be either autopolyploids or allopolyploids of parents having the same chromosome numbers.

If the chromosome numbers show an irregular progression (e.g., n=6, n=8, n=15, n=21), polyploidy may be involved through allopolyploidy involving parents having different chromosome numbers. In the series cited, the n=21 would presumably be an allopolyploid of the $n = 6 \times n = 15$; the $n = 15$ would presumably be an allopolyploid of either $n = 6$ and an undiscovered species with $n = 9$ or of $n = 8$ and an undiscovered species with $n = 7$. The intricacies are well illustrated by Fig. 23 in Chapter 5 (p. 110). Because allopolyploid parents may not be closely related, an extensive knowledge of the cytology of species outside the immediate study complex is essential to understanding possible and/or probable hybrid combinations leading to particular instances of allopolyploidy.

If these circumstances occurred between plant sets, polyploidy would almost certainly be involved. Tests with colchisine on lower-numbered sets or hybrids between the sets might give proof of their polyploid origin (see p. 107). If they occurred in animals, where demonstrated polyploidy is rare, there would be serious doubt concerning their polyploid nature.

If two polyploid sets of organisms proved to be at the same ploidy level and were cytogenetically similar, they could be sister species that had arisen from a common polyploid ancestor. Thus if in set A, n=6; in set B, n=12; and in set C, n=12; then sets B and C could have evolved from a parental form having n=12 that arose by polyploidy from set A. Sets B and C would then need to be tested as isoploid species, as outlined in the following section.

Inter-set Hybridization

If we have living material of bisexual sets that will reproduce successfully in the laboratory or garden, it is usually possible to arrive at a highly probable answer to the question, Do two sets represent one or two species? The ultimate goal of the question is to determine if any two sets of individuals represent lineages that are evolving independently in nature. The procedure followed is first to cross the two and observe the production and behavior of hybrid progeny, then to check the results against circumstances observed in the field.

In the case of monoecious flowering plants, it is important to know if the species can be self-pollinated or if they must be cross-pollinated. In the case of many animal species, difficulty is often encountered in obtaining laboratory mating. Many ingenious methods for accomplishing this have been devised, which can be found in the literature pertaining to individual groups.

In these tests, individuals of one set are crossed with those of another, and any progeny recorded. Individuals of the F_1 generation are crossed, and any resulting F_2 generation individuals likewise crossed. Backcrosses are made between F_1 and F_2 individuals and the two parents. If three or more sets of individuals have been segregated, each is crossed in like fashion with every other set. The ideal is to make several crosses between each two sets, representing a grid from various localities.

Controls should be run by crossing individuals from within the same sets. If most of these controls produce few or no progeny, then the techniques being used are obviously inadequate and negative results from inter-set crosses would have no significance. The analysis given below is based on the premise that only well-tested rearing, culturing, or growing techniques have been employed, as stressed by Usinger (1966).

The results of crossing any two sets of organisms would fall within one of three categories, each having its peculiar implications.

1. *No progeny produced or the F_1 generation sterile.* The two sets would be genetically isolated species, corresponding to step 4 in Fig. 16 in Chapter 5; if sexual or other isolating mechanisms were involved, the correspondence would be to step 5. Most distinctive animal species and many plant species typify this category.

2. *Markedly reduced fertility in some or all tests.* To this category belong the great majority of instances in which it is difficult to determine whether two sets of individuals represent evolutionarily distinct lineages.

If the fertility were reduced in all tests, the implication would be that the two sets represented two species that had reached step 4 in Fig. 16 in chapter 5, but which had not yet evolved sexual or other rigid types of reproductive isolation. Thus many species of *Drosophila* hybridize to a considerable extent in the laboratory (Patterson and Stone, 1952), and large numbers of plant species hybridize to some extent in the garden (Stebbins, 1950). It was the consensus in both references that allopatric, morphologically distinct species produced hybrid progeny much more frequently than did sympatric ones. The inference is that if the ranges of sister species should come in contact, the allopatric entities would rapidly develop isolating mechanisms of one sort or another, reducing hybridization between them. In sympatric species that will hybridize in the laboratory or garden, few hybrids of many combinations are found in nature, indicating that reinforcing isolating mechanisms are present in the field that are nonoperative in the laboratory, essentially implying that in such cases speciation is well on the way to step 5 in Fig. 16 in Chapter 5. It would seem that the more the experimental fertility is reduced or the fewer the number of natural hybrids found over the whole range, the more likely it is that the two sets of specimens are species in an evolutionary sense.

If the fertility was high in some tests, low in others, then certain alternatives would be suggested. If one set occurs at one end of the total range, the other at the other end, with intergrading character states in between, the two categories are probably variants of one species. Presumably the two character states would be correlated with different ecological factors occurring across the range. In such a situation, populations from distant points of the range might exhibit considerable hybrid sterility, but those from closer points would be highly fertile. The population-to-population fertility would link the entire species into a single interbreeding unit, even though genetic change at distant parts of the range would spread only slowly to other parts. Examples include the fruit fly *Drosophila pseudoobscura* in which the vigor of the F_2 generation decreases when populations from California are crossed with those from Nevada, Utah, and Colorado, respectively.

In the herring gull *Larus argentatus* this phenomenon has resulted in a strange situation. The range extends around the Arctic Ocean and the two ends of the range, represented by different subspecies, meet and overlap in northwestern Europe. The two overlapping forms do not hybridize but live together like separate species although connected genetically by a ring of intermediate interbreeding populations. This peculiar *ring species* undoubtedly started out possibly in northern North America as a limited linear species in which the gene flow throughout the population was sufficiently small that the end populations became more and more incompatible genetically. Presumably the end populations spread east and west around the subpolar regions and by the time they came in contact in Europe they were completely isolated genetically although there apparently is a genetic continuity throughout the ring as a whole between neighbor-to-neighbor populations.

3. *Abundant fertile progeny produced in most or all tests.* In most instances, the two sets would be variants of one species. If the sets were allopatric, then the different character states would be allelic substitutions that had evolved in isolation, exemplifying step 3 in Fig. 16 in Chapter 5. If the sets were sympatric, then the character states defining the sets would probably be few in number and linked in an inversion on the same chromosome. Otherwise the character states would have segregated more randomly as in Table 6.1

There is one circumstance in which this outcome of successful experimental crossing would occur with the existence of more than one species. Many distinctive plant species will hybridize successfully in the garden, but remain distinct in nature because of various isolating mechanisms (see p. 84). For example, four California species of the section Peltanthera of *Penstemon* having variously overlapping ranges remain distinct in nature although they are interfertile in all directions. Straw (1956) discovered that each of the three species whose ranges overlap the most are pollinated by different animals, one by a hummingbird, one by a large bee, the third by a smaller bee, and that in each species the flowers are adapted to their principal pollinator. So effective is the disruptive selection of the differential pollination that Straw found only scattered hybrids where any two species were sympatric.

Turning to a different type of example, the ironweed genus *Vernonia* has a number of sympatric species in eastern North America that interbreed freely in the garden, but remain distinct in nature. Occasionally large hybrid swarms occur in disturbed areas to which the hybrids are presumably better adapted than either parent. The inference is that hybrids are produced continuously, but in the parents' ecological settings they are outcompeted by the parental types (Jones, 1972). In plants it is therefore necessary to see if the hybrid types produced in the garden also occur in the same proportion generally in the natural range. If so, one species is indicated. If the hybrid types are absent or greatly reduced in the native range, then each category represents a different species. It is possible that this phenomenon occurs also in animals, especially in groups that are host specific, but the subject has been little studied. The hybrid swarms of Mexican towhees analyzed by Sibley (1954) and Sibley and West (1958) may be evidence of this phenomenon.

Ecological Correlations

Circumstances concerning allopatric sets outlined above apply also to sets associated with certain ecological conditions. If one set of specimens is associated with one host, another set with another host, crosses between the two sets can be attempted using the mother's and the father's hosts as a base of operation in duplicate experiments. The results will either give a value of the genetic incompatibility (if any) between the two sets or indicate whether or not the F_1 progeny can survive on a parental host. Host transfers can be made, which will indicate the ability or inability of the sets to reproduce on the "wrong" host.

A second ecological isolating factor concerns time. If one set has its reproductive period at a completely different time from the other, for example, one mating in August–September, the other in April–May, or one blooming in June, the other in September, the two are reproductively isolated. To test the genetic incompatibility of the two sets, it would be necessary to manipulate one of them so that its reproductive period was the same as the other. Chapman and Jones (1971) did this in achieving hybrids between two *Senecio* species, "forcing" one and giving the other cold treatments to slow its growth.

If in crossing tests the two sets proved to be sexually isolated or their progeny were either sterile or had markedly reduced fertility, the two sets would be indubitably distinct species representing steps 4 or 5 in Fig. 16 in Chapter 5. If fertility were higher, the two sets might be considered as representing only step 3 in Fig. 16 in Chapter 5 and designated accordingly.

Intra-set Hybridization

Each set should be examined to determine if it represents a cohesive single unit. Individuals from different parts of the range should be crossed; if the set occurred on different hosts, individuals from different hosts should be crossed; and so on. If any subsets prove to be genetically incompatible with other members of the set, these subsets would be cryptic species.

Giving species names to cryptic species raises problems with regard to classification. Traditionally, species have been considered entities that could be keyed out in floras or faunas on morphological grounds. Cryptic species cannot be treated in this "neat and tidy" fashion, but for adequate identification may require song recordings, chromosome data, breeding with tester sets, or other nonmorphological information. A simple way to solve this problem is to afford cryptic species their evolutionary status by recognizing them as species, but to key them out to morphological units designated as complexes or superspecies. At present, however, there is no consensus as to this kind of treatment although one is needed badly.

Cryptic species have now been found in so many plant and animal groups that no species untested for them can fail to be suspect of harboring them. The total known number is not large, but the distribution among various taxa is extensive (see p. 97).

Perspectives

It should be clear at this point that in many instances it is impossible to give a hard-and-fast answer to the question: Do these sets of individuals represent one or more than one species?

Apomictic and polyploid species arise by the occurrence of a single mutation (sometimes following hybridization), and therefore adequate cytogenetic and/or reproductive information makes the status of these entities clear.

In those species arising by gradual mechanisms there is an evolution from an initial splitting of a parental species into genetically isolated populations, and the gradual evolution of these populations to a condition of rigid genetic and sexual isolation. The result is two or more species where before there was only one. Because different lineages could split at different times and could differ in the rate of becoming genetically distinct, it follows that at any one time we should expect at least some lineages to be in various stages of one of the gradual mechanisms (Fig. 16) in Chapter 5. It is this circumstance that sometimes makes it difficult for us to decide whether two sets of isoploid, bisexually reproducing organisms represent one species or two.

Instead of striving for a dogmatic "Yes" or "No" answer, we should assess data gathered according to the preceding tests and, in the light of principles known concerning the speciation processes outlined in Chapter 5, we should then be able to reach a probability answer. For example, we might conclude either that sets 1 and 2 are probably one species or that they are probably two species. In either event our conclusion should incorporate the data we assembed, the reasoning we used, and our judgment as to where in the evolution of contemporaneous species the entities under study belong.

SPECIES ANALYSIS WITHOUT EXPERIMENT

The preceding analysis of species was based on techniques and tests requiring living material that could be crossed and studied cytogenetically. Only a very small fraction of the known species of the world's biota meets these requirements. The status of possibly 99% of the present day species and 100% of the known fossil species must be based on a far smaller increment of information. For many species, living material may be available but either time or known techniques are not available to perform genetic experiments. For example, many large trees require ten or twenty years before they reproduce; it would take half a century to test the reproductive behavior through the F_1, F_2, and F_3 generations necessary to establish possible genetic breakdown. Large animals have similar examples. In these instances, human time and resources have been insufficient to carry out extensive investigations.

In many groups of animals for which living material is available, techniques have not yet been devised for inducing mating or for rearing the young under laboratory conditions. This problem is especially troublesome in many groups of insects. Periodically new breakthroughs are made, as in the case of certain mosquitoes. It had long been impossible to induce laboratory mating of many groups, until McDaniel and Horsfall (1957) devised a successful method using CO_2^- anaesthetized females and decapitated males.

The great bulk of the world's two million or more species are known to us only on the basis of preserved specimens in collections, and this is true also of most of the sets of specimens that we might suspect to be previously unknown species. For some entities, abundant material is available from a large number of diverse localities; for others, lesser amounts are available; and we finally reach the point at which we have only one specimen per suspected entity.

In view of the unusually large number of species or entities for which we do not have experimental evidence used earlier in the chapter for inferring their specific status, it is necessary to develop other criteria for making these inferences. The criteria that follow are based on the occurrence of different character states in collected specimens and the geographic and ecological distribution of these. So many character and distributional variables are known that it seems impossible to anticipate every combination. Nevertheless, guidelines can be formulated that accommodate the combinations usually encountered and that may serve as stepping stones to additional procedures.

The premise used is that the differing character states of sets of individuals are usually an indication of the genetic similarity or dissimilarity of the sets. We assume that such differences are almost always related to the differential action of natural selection regarding various parameters of the environment and, in the case of sympatric species, that they also represent the evolution of interspecific isolating mechanisms. Thus two sympatric sets that differ in several striking morphological traits such as those in Table 2 would appear to be sufficiently different genetically to make it a safe assumption the two sets represent separate breeding populations. Sets that differ by only one or two small differences might represent a single interbreeding unit in which the differences were simply allelic expressions of a heterozygous genome. The above premise leads to the following obvious limitations:

Cryptic species. Any one set may contain two or more cryptic species that would not be brought to light until intra-set hybridization was possible. For example, to date no character differences are known to identify the cryptic species within the *Drosophila paulistorum* complex (the semispecies of Dobzhansky and Pavlova); they can be identified only by mating to individuals belonging to a set of standard cultures. According to character differences, all species of the *D. paulistorum* complex would sort to a single set or species, as they were previously considered before intra-set matings were made. The same circumstances apply to the *Cyclops vernalis* complex (see p. 97). Thus the guidelines given below would not identify cryptic species.

Cytogenetic information. It is impossible to obtain cytogenetic information from material previously collected according to most techniques. This includes dried herbarium specimens, bird and mammal skins, pinned insects, and material preserved in formalin or alcohol. It is therefore impossible to identify polyploidy, hybrids, or other cytogenetic phenomena by direct chromosomal study in the bulk of the museum material available. Notable exceptions are pollen grains and epithelial cells that are often larger in polyploid plants.

GUIDELINE PROCEDURES

These procedures are designed for use with preserved material only. At this point the investigator has delineated two or more sets of specimens characterized by distinctive states of one or more characters. A simplified and common situation is one involving either two distinctive sets, or two distinctive sets plus a third set that represents intermediates between the two distinctive ones. The problem is to determine whether the sets represent one or more species.

Apomictic vs. Bisexual Types

As when treating experimental material, the first step is to test for type of reproduction, primarily to find out if either or all sets are apomictic or bisexual. This distinction is often difficult to make on the basis of preserved specimens. If random intermediates occur between the two sets, then some interbreeding is evident, ruling out the possibility of apomixis. If not, then further tests for apomixis are in order. In flowering plants, apomictic species may be suspected if the pollen is always shriveled or the seeds shriveled or aborted. In dioecious species the absence of males may indicate apomixis. Even if all collections of a set indicate apomixis, it is nevertheless difficult to be sure that the set does represent an apomictic species. Apomictic plants often have a series of forms connecting the sexual lines in a complex fashion; deciphering these situations usually requires living material.

It is well known that under certain climatic conditions a goodly number of normally bisexual plant species do not set seed but reproduce only vegetatively (Stebbins, 1950). Certain insects do likewise. In many insects, suspected apomixis has been demonstrated to be due to peculiarities of collecting, especially when the males live only a few days and the females for several months. As a result, great caution should be used in deciding that a set of preserved specimens are apomicts.

If it appears certain that both sets are apomicts, the two will need to be classified on the merits of their differences. If one set is an apomict, the other bisexual, these also will need to be classified on the same basis. If both are bisexual, or if the evidence for not considering them so is inconclusive, the investigation proceeds to the following steps.

Bisexual Species

In a large number of investigations concerning contemporaneous species, the problem resolves itself into deciding the species status of two sets of individuals, sometimes with a set of intermediates. This simple situation is usually expressed in a limited number of geographic character state distributions that can be used to infer the probability of the existence of one or two species. Figure 1 diagrams procedures for testing these probabilities for two sets, 1 and 2, and possible intermediates, 1 x 2, having a simple distribution. More complex patterns are illustrated and discussed in a later section.

The first step is to plot the distribution of sets 1 and 2 and any intermediates 1 x 2, on a map, then to follow the flow lines in Fig. 1 until the end result is reached. This result will be either a number, referring to one of the following explanatory notes giving the probable answer (s) to the problem or a reference to the later section discussing more complex distributions.

Explanatory Notes

Sympatric distributions. (Fig. 2a)

Note 1. Intermediates 1 x 2 absent. If sets 1 and 2 differ in two or more character states, they are probably two species. The larger the number of differences, the greater is this probability. If the difference is in only one character state, two species might be involved, but there is a possibility that the differences represent two alleles, one exhibiting complete dominance.

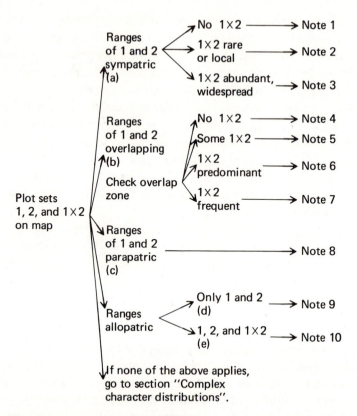

Fig. 6.1 Flowchart of character-range comparisons used in guideline procedures for species analysis. Designations (a) through (e) refer to those parts of Fig. 2.

Note 2. Intermediates 1 x 2 local. This situation, common in plants, indicates two species with occasional hybrid swarms. The intermediates are often associated with a disturbed or unusual habitat. Thus in Illinois, hybrid swarms between *Quercus marilandica* and *Q. palustris* occur only in certain local areas. Hybrid swarms have been listed for several vertebrate groups (Miller, 1939; Rudd, 1955; Hubbs, 1955). In Massachusetts, Hanson (1960) discovered hybrid swarms between the aquatic winter stoneflies *Allocapnia maria* and *A. minima.* The former is a slow brook species, the latter a rapid river species, hence the two are usually separated in any one locality by stretches of an intermediate type of stream in which neither species lives. But in some instances where a brook flows directly into a large river, the two come in contact and hybridize. Except at these hybridizing sites, the two species exhibit no evidence of introgression. Presumably barriers to interbreeding between the parent species break down under some peculiarity of a locality, and local hybridization follows.

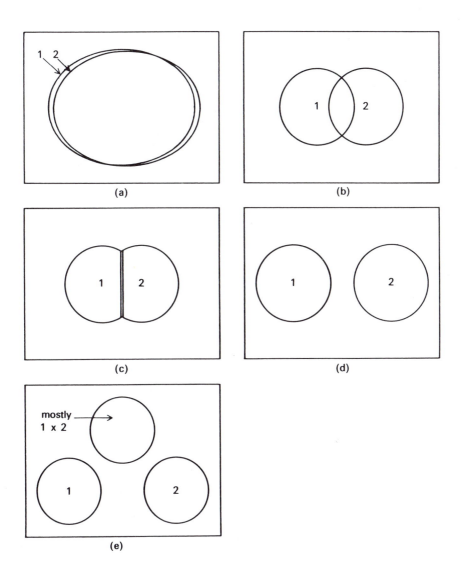

Fig. 6.2 Common distribution patterns displayed by two sets of individuals.

Note 3. Intermediates widespread and abundant. If the various character states are distributed randomly in the intermediates, then only one variable species would be present. If the intermediates comprised a nonrandom set of individuals, each having the same combination of intermediate character states, the intermediate set would presumably be one species and sets 1 and 2 would each be a separate species. This situation occurs commonly in insects.

Overlapping distributions. (Fig. 2b)

Note 4. No intermediates in overlap zone. Two species are indicated. Presumably sets 1 and 2 evolved in previous isolation to the point of being completely reproductively isolated. When their ranges subsequently overlapped, no effective hybridization occurred.

Note 5. A few intermediates in overlap zone. Presumably in this instance also, sets 1 and 2 are separate species, but did not quite reach complete reproductive isolation before the two ranges came into contact, resulting in infrequent hybridization.

Note 6. Intermediates predominant in overlap zone. There is every indication that only one species is involved. Previously sets 1 and 2 evolved character state differences in isolation but no reproductive isolation was involved (step 3 in Fig. 16 in chapter 5). When their ranges came together the two populations interbred and began the formation of a new variable population. Presumably the range of overlap will increase and result in the widespread variable species discussed in Note 3, above.

Note 7. Intermediates frequent in overlap zone. This combination is one of the most puzzling to decipher. It is intermediate between the situation discussed in Note 5, which indicates that sets 1 and 2 are distinct species, and that in Note 6, which indicates that the two sets and their intermediates form only one variable species. If the latter were true and if interbreeding and reproductive success were random, a plot of character states would produce a unimodal distribution according to the Hardy-Weinberg law. On the other hand, if there were selective forces, even small, acting against either the fertility or success of the hybrids, a plot of the character states in the overlap area would show a bimodal distribution. Theoretically, the more pronounced the bimodality, the greater would be the probability that the two sets were distinct species between which reproductive barriers were evolving. Greatly pronounced bimodality would represent an early stage of step 5, Fig. 16 in chapter 5. Slight bimodality might be explained because of certain character states having no phenotypic intermediate expression, such as single-allele dominants.

Parapatric distributions (Fig. 2c)

Note 8. Ranges of the two sets abutting, with an extremely narrow overlap zone. The two sets are indubitably two species. The sharp boundary between them is due to either genetic or ecological reasons. In European hares of the genus *Lepus* (Notini, 1941, 1948) and Australian grasshoppers of the genus *Moraba* (White et al., 1964), the parapatric distribution is maintained by a combination of three factors: (1) very low vagility (the grasshoppers are wingless), (2) free intermating between the two entities, and (3) hybrids that are sterile. The wingless Eurasian young-mare grasshopper *Podisma pedestris* may be a similar example. John and Hewitt (1970) and Hewitt and John (1972) report a chromosomal race of *Podisma pedestris* in the southern Franco-Italian alps that may eventually prove to be a cryptic, genetically incompatible sister species of the widespread form. In two European grain stem sawflies imported into eastern North America, the parapatric distribution is maintained by competition (Udine, 1941). The northern species outcompetes the southern one, which latter is restricted to a southern area in which the northern species cannot survive.

Allopatric distributions (Fig. 2d,e)

Note 9. Ranges of each set separate, no intermediates. The lack of intermediates would indicate little or no genetic contact between the two populations, and differences sufficient to delineate them as two sets would indicate that the two had undergone independent evolution.

 If the isolated populations are each represented by large numbers of individuals, and if their character ranges do not overlap, they are probably distinct genetically as well as with respect to character. This need not, however, be universally true, as is obvious from examples in note 9, above. If only modally distinct, they would seem to be the same species. But there is always the possibility that such isolated populations are nevertheless genetically incompatible. Porter (1969) crossed allopatric populations thought to be the same frog species *Rana sylvatica* and found them genetically incompatible to a high degree and therefore distinct species. Under these diverse circumstances, the following are reasonable operational judgments. If allopatric populations of a complex differ in character as much as related sympatric species in the same taxonomic group, they should be considered species until demonstrated otherwise. Distinguishable allopatric sister populations but with differences of a lesser degree should be given some sort of epithet or designation; although this latter procedure is controversial, it does serve to keep attention drawn to what might be a knotty but biologically interesting problem.

Note 10. Sets 1 and 2 each with a separate range, a third allopatric population with 1, 2, and intermediates. (Fig. 2e). This type of distribution indicates the existence of one species with a special history. It is almost certain that the parent of sets 1 and 2 was fragmented and the two isolated populations evolved character differences but no reproductive isolation, resulting in sets 1 and 2. Later, with changing climates, set 1 and set 2 dispersed into the area now occupied by the intermediates, and when they came into contact hybridized and formed a variable population. The present allopatric ranges of sets 1 and 2 indicate further that climatic conditions continued to change and intervening populations did not persist. Sets 1 and 2 would therefore be relict populations living in local areas having special ecological conditions. In North America common examples include relict populations on higher mountain elevations south of the main range, or in spring-fed streams south of the area occupied by the intermediates (Ross and Ricker, 1971). Western alpine plants have similar examples.

COMPLEX CHARACTER DISTRIBUTIONS

The preceding distributions of sets of individuals representing different character states include the most frequently encountered situations. Other known distributions may be more complex or puzzling. A brief sampling of some of these follows.

More Than Two Sets.

If the study group falls into three or more sets of specimens, these may be analyzed initially by following the flowchart of procedural steps outlined in Fig. 1 and the appropriate explanatory notes. If three sets are involved, and their possible intermediates, each must be compared with each of the other two, because either one, two, or three species might be involved.

A marked deviation from the two-set situation arises when the three or more sets have overlapping ranges. These are diagrammed in Fig. 3, using a model of three sets. The overlaps may be tandem (Fig. 3a) or form a triple overlap (Fig. 3b). Each tandem overlap may be treated separately according to notes 4 to 7. If few or no intermediates are present, the triple overlap will resolve itself according to notes 4 and 5; that is, three species will be indicated. If intermediates are frequent or abundant between all three sets, the intermediates form a chaotic combination of characters states. For example, in the western North American sawfly *Tenthredo pectoralis* the populations in Alaska, the Pacific coast states, and the more southern eastern ranges of the Rocky Mountains differ from each other in color patterns of the thorax, abdomen, and legs; within each area, they are relatively uniform. In Montana and immediately surrounding areas the populations are variable in the extreme, representing almost every conceivable combination of the various states of all three characters. In this example it seems obvious that only one species is involved, which represents the result of post-Pleistocene dispersals of populations isolated during the Wisconsinan. The complexity of the intermediates, however, raises the question as to how to detect and interpret various stages of bimodality should they occur.

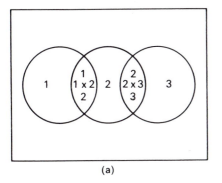
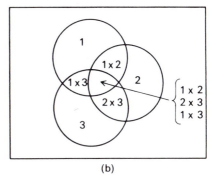

Fig. 6.3 Two types of overlapping ranges displayed by three sets of individuals.

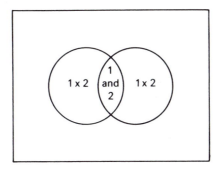

Fig. 6.4 Apparent (but not real) distribution of character states exhibited by potential cases of character displacement. For explanation, see text.

Character Displacement

A distribution that at first seems puzzling is shown in Fig. 4, in which the two sets occur together in a central area and the intermediates in the two peripheral areas. Such a distribution is almost certain to be an example of character displacement, explained in detail on p. 76. On closer inspection, the two peripheral "1 x 2" populations will undoubtedly prove to be slightly different, and sets 1 and 2 in the central zone will represent exaggerations of these differences. Two species would be involved.

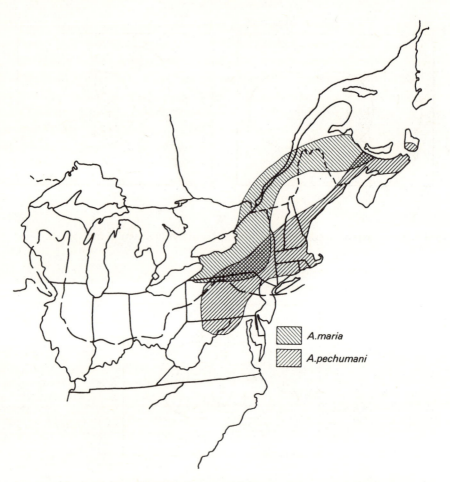

Fig. 6.5 Ranges of *Allocapnia pechumani* and *A. maria*, with hybridization evident only in northern zone of overlap. The broken line is the approximate southern limit of the Wisconsinan glaciation. For discussion, see text. (Adapted from H. H. Ross and W. E. Ricker, "The classification, evolution, and dispersal of the winter stonefly genus *Allocapnia*, *Ill. Biol. Monog.*, 45:1–166, 1971.)

Differential Hybrid Indications

In some instances two sets are sympatric with no intermediates in one part of their range, but intermediates occur in another. An example is a pair of North American winter stoneflies, *Allocapnia maria* and *A. pechumani*. Their known ranges (Fig. 5) overlap broadly in New York and Pennsylvania and in a smaller area in New Brunswick. No intermediates have been found in the southern overlap zone, but they are common in the northern one. No bimodality has been detected, but each collection made thus far is predominantly one species or the other. This circumstance and the narrowness of this northern overlap zone suggests hybrid sterility with some backcrossing. The two entities seem to be distinct species formerly displaced south of the glacial maximum border, perhaps one species on each side of the Appalachian Mountains. During the early stages of the last glacial dissipation, the southern populations of the two genetically incompatible entities came into contact through the low passes and have since evolved rigid reproductive isolation. The northern populations spread to the northeast as the ice disappeared, *A. pechumani* to the west, *A. maria* to the east of the higher northeastern mountains. The two finally came into contact comparatively recently in New Brunswick and have not yet had time to evolve complete reproductive isolation. These postulated events fit the known geological history of the area and the phylogeny of the genus (Ross and Ricker, 1971).

Hybridization between the tiger beetles *Cicindela duodecimpunctata* and *C. oregona* in western Canada seems to follow this pattern (Freitag, 1965). As in the *Allocapnia* example, only the most northern populations of the two species of *Cicindela* show evidence of extensive hybridization; indications of hybridization in more southern populations are typical of introgression.

Similar circumstances have been observed in mammals, especially with the subspecies of *Peromyscus maniculatus,* some of which behave like species on one side of their range and yet intergrade with their relatives on the other.

Migratory Birds

One of the most puzzling circumstances in deciphering problems of contemporaneous species is concerned with what are called subspecies, races, or populations of migratory birds. In well-studied examples such as the Canada geese, many of the entities are such tightly knit groups, keeping together as discrete populations at all times of year, that they appear to be evolutionarily distinct lineages (Fig. 6). There is evidence of only a low percentage of hybridization between adjacent populations, and abundant evidence that the same birds return to the same northern nesting localities (Hanson and Smith, 1950). There is a real question as to the evolutionary status of these and many other populations or entities of migratory birds.

Conclusions

These examples of complex character distributions are but a few of many that could be assembled. The purpose in citing them is to give a brief insight into the variety of circumstances that might be encountered in studies of speciation, hints as to some of the reasoning that might be helpful in their solution, and some of the types of problems for which we have as yet no ready answers. Perhaps the ultimate purpose in presenting these examples is to introduce the thought that the investigator may encounter intriguing and novel

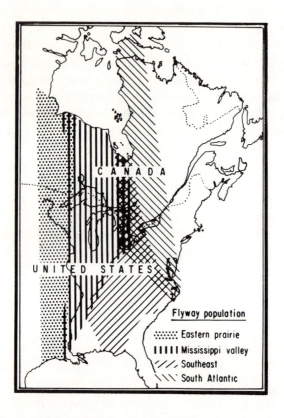

Fig. 6.6 Map showing roughly the main ranges of the four populations of Canada geese nesting in the Hudson-James Bay region. (Adapted from Hanson and Smith, in H. H. Ross, *A Synthesis of Evolutionary Theory,* © 1962. Reprinted by permission of Prentice-Hall, Inc., Englewood Cliffs, N. J.)

situations in the study of contemporaneous speciation that will require new approaches for their solution.

SCANTY MATERIALS

A recurring problem in systematic studies is the disposition of species suspected to be new to science for which only one or a few specimens are available. Opinions on the question differ widely. Some believe that the specimens should be set aside until more material is available. Others believe that every specimen should be identified to species, and that uniques should be named. It seems to me that a judgment might differ under various circumstances.

If the few specimens in question are most closely related to a species for which an abundance of material is known, the variational limits of the latter can be determined accurately. If the suspect specimens are well beyond these limits, there is an excellent chance that they do represent a distinctive species and should be named. If just off the limits of their relatives, haste in naming would ordinarily not be indicated. However, if the investigator were finishing a revisional or monographic work or an important regional treatment, then it is desirable to give an unambiguous species epithet to every specimen showing diagnostic features.

In these instances there is one merit to being a splitter. A proposed name that proves to be a synonym can readily be assigned to its proper place and the correction is forever clear. On the other hand, if the specimens are assigned to and recorded as a related species, but later prove to be different, it is difficult to erase the misidentification from the continuing literature.

APPLICATION OF NAMES

In the light of considering contemporaneous species as representatives of evolutionarily distinct lineages, it seems desirable to equate the entities we call species with the concept of evolutionary distinctness. As pointed out in Chapter 5, naming sequential and apomict species poses certain subjective decisions based solely on amount of character or ecological differences between the various entities that can be recognized.

Polyploids constitute a special case. Allopolyploids are almost universally designated as species, which is logical because there is little or no genetic exchange between them and their lowerploid ancestors. Autopolyploids also have less genetic interchange than many recognized isoploid species such as those of *Vernonia,* yet there is reluctance to recognize them as species because of the frequent difficulty in recognizing them without chromosome counts. The question arises as to whether this convention of naming does not obscure the evolutionary realities of the situation.

With bisexual isoploid species there is often doubt as to whether or not two discernible entities represent evolutionarily distinct species. A number of procedures and criteria have been outlined above to test these possibilities between isoploid species, based on theoretical considerations developed in Chapter 5. For isoploid bisexual species, the following suggestions are made, in particular reference to Fig. 16 of Chapter 5:

1. The term species be used for entities that give evidence of belonging to steps 4 or 5.

2. The term subspecies be used for populations or entities that seem to be almost species (upper level Step 3) and that give evidence of past evolutionary events.

3. Some non-Latinized designation be used for all other populations or entities at an earlier stage of divergence. Locality or range is useful for designating geographic entities. Geneticists use localities or letter designaions such as WT, MD, KL, etc., for gene arrangements on chromosomes. This practice might be applied to geographic varieties (especially ecotypes), strains, or character combinations. Genetic mutants and varieties are designated in a number of ways.

The use of the category subspecies was criticized by Wilson and Brown (1953), defended by Smith and White (1956), and argued by many others. Subspecies are nevertheless employed by systematists in many groups, usually in the connotation given above. Criteria for delineating subspecies vary from group to group, and it is advisable to study the practices followied in the particular group under study in an effort to correlate the current practices with evolutionary implications. Mayr (1969) gave an excellent summary of many of these practices used in animal groups, Grant (1971) for plant groups.

7

Phylogeny

The first objective in biosystematics includes determining as nearly as possible how many species there are or have been. The second objective is to figure out the phylogenetic relationships or genealogy of these species. Methods for attacking the first objective were given in the previous two chapters. Investigating the second is the concern of this chapter.

THE RELATION BETWEEN SPECIES AND LINEAGES

We can't actually see lineages progressing through time because the 3000-year human period of detailed biological observation is barely a momentary glimpse compared with the 3 billion or more years that life has been evolving. But species, which are known to us as samples of a cross-section of a lineage, give us glimpses of the lineages at different points in time. From these glimpses we try to reconstruct the entire family tree of life.

It is obvious that the more species we can investigate, the more complete will be our family tree. Because we can investigate only about one percent of the probable existing species from genetic and behavioral information, most present-day phylogenetic studies must be based on species known only from preserved material. Hence our preoccupation in the last chapter with procedures that will give probable answers to questions about speciation. These techniques extend our horizon and perspective possibly a hundredfold as far as understanding the taxonomic diversity of the universe.

The need for knowing this diversity in order to further phylogenetic study emphasizes two points about which there is considerable prejudice. The first concerns the reticence of many systematists to describe new species on the basis of one or a few specimens. Yet if it represents a distinctive morphological entity, one preserved specimen gives us a remarkable amount of information about an entire phylogenetic line of which we had no previous knowledge. The second point concerns subspecies. If named, they give us notice of incipient species that might be important in our phylogenetic thinking, but would be missed by the phylogenist if not given a name.

In their hypothetical framework, speciation and phylogeny have much in common. Both use the scientific method with its implication of probability rather than certainty in expressing their results. Both use comparative characteristics as the basis of arriving at their results. But there the similarity ends.

Speciation uses data from experimental genetics, comparative behavior, biochemistry, and the geographic or ecological distribution of character states as an expression of genetic incompatibility or sexual isolation. Even in a well-documented fossil assemblage of related species the data permit an assessment of some of these traits.

The minute we get into phylogeny we are in the realm of the *ancestors* of known species, living or fossil. Certain species we have identified may be like or almost like some of these ancestors, but this situation is not apparent until it can be inferred. Because forms ancestral to living species plus all fossil species are not represented by living specimens, they can be known only as morphological entities. We can therefore relate them to living species only through morphological comparisons. Phylogeny is thus primarily concerned with comparative morphology. Geographic and ecological hypotheses must be adduced after morphology establishes the phylogeny.

GENERAL ASSUMPTIONS

It can be said with some justification that speciation is an extension of genetic theory. In this same vein, phylogeny is a corollary of the theory of organic evolution. In attempting to draw principles (general truths) or operational methods of assistance in unraveling the phylogeny of a particular group, we must use case histories or phylogenies that are well documented and intensively studied as our sources. The best documented case histories involve: the vertebrates whose bony structures fossilize readily, the higher plants whose structures also fossilize readily, and certain marine protozoans and invertebrates whose remains fossilize not only readily but in large numbers.

On these bases, we can outline the following assumptions or premises of phylogeny.

1. Phylogeny assumes that the theory of organic evolution is correct. If so, there is a correct family tree for all groups, and this family tree will express the course of evolution, whether it be convergent, divergent, dichotomous, polychotomous, reticulate, or broken. Broken segments would represent lineage extinctions.

2. Because these events happened in the past, we cannot observe them but must infer them from any available evidence. In this light we can define the *phylogeny* of a particular group as *our concept of the path of its evolution.*

In this context, phylogeny is a hypothetical construct that we hope will approximate the facts of the universe; that is, how various members of the group actually did evolve. Individual phylogenies are therefore not dogmatic statements but scientific inferences subject to negation or improvement in the light of new evidence or ideas.

3. Sufficient fossil evidence has been accumulated to provide many highly probable examples of the evolution of specific taxa through geologic time. These examples include groups which have a long history but became extinct eons ago (e.g., the Fusulina, C. A. Ross, 1969), groups which were abundant in the past but now rare (e.g., the echinoderm class Crinoidea, Moore and Laudon, 1943; and the elephants, Lull, 1917), and groups which were abundant in the past and continue to be so in the present (e.g., the plant genus *Ginkgo* Tralau, 1968). These examples give us case histories from which to extrapolate additional assumptions.

4. Species in the past resembled those of the present in the amount of intraspecific variation. This premise was established by Simpson (1944) on the basis of morphological criteria, but it presumably applies also to genetic composition, polyploid and apomict potentials, ecological tolerances, etc. This concept is an extension of Hutton's Law of Uniformitarianism, devised for geological events but undoubtedly true for many biological events also.

5. Each phylogenetic line is one species thick. This is simply an extension of assumption 4, above. It implies that no past phylogenetic line possessed a greater store of genetic variability than we can detect experimentally in the more variable species existing today.

6. The more complex the structure, the less likely that it would evolve independently twice. By way of examples are the intricate sifting mouth structures of the baleen whale, the highly evolved "seed" of angiosperm plants, and the highly complex insect eye, each of which apparently evolved only once. Whereas, on the other hand, we note the evolution of jumping hind legs in such distantly related animals as grasshoppers and kangaroos, and the evolution of compound leaves in relatively unrelated plants such as ferns and hickories. In both examples, the parallelisms appear to involve relatively small genetic changes.

7. Complex structures once lost are never regained in the same form. There is evidence that simple structures lost in evolution may sometimes be regained. In insects, certain primitive cross-veins are lost in entire subordinal branches, then reappear in one or two highly evolved families. The juxtaposition of the "recovered" cross-veins is so remarkably like the primeval type as to warrant the assumption of a regained character. Several mechanisms have been suggested to account for this phenomenon, particularly the temporary masking of

characters by shifts in heterochromatin (Brink, 1960, 1964). Such shifts
have been invoked to explain the reappearance of only simple characters.
There is no need to invoke them for explaining the reappearance of complex
characters because no examples of the latter phenomenon are known. There
is no evidence of a mammal lineage losing its ear and subsequent daughter lineages regaining it, or of plants completely losing their photosynthetic activity
and having this activity evolve again in a continuation of the lineage.

Certain complex structures *seem* to be lost and then regained. There is
excellent paleontological evidence that the turtles lost and regained their external armature several times. However, in each reappearance the armature
was of a type different from that possessed before. A somewhat different
type of loss and reappearance involves the fins of fishes and porpoises. In this
sequence, the fish fin evolved into a typical land vertebrate limb; then this in
turn evolved into a porpoise fin. Although superficially alike in appearance,
the fish and porpoise fins are dissimilar in structure.

8. Characters are lost more frequently than new ones evolve. An outstand-
ing example concerns the wings of insects. In hundreds of lineages, many not
closely related, the change in basic venation proceeds by the loss of one vein,
then another, and so on until the wings have a greatly reduced venation. Such
wing losses are usually correlated with reductions in absolute size of the organism. In the Lepidoptera, Michener (1949) tabulated apparently lost characters
vs. apparently newly evolved ones, and found that losses outnumbered novelties
ten to one. In plants, a numerical reduction in number of stamens has occurred
in many lineages, reminiscent of the parallel losses in insect wing venation.

9. Different characters may evolve at different rates in the same lineage.
For example, in many insect lineages the male genitalia have evolved elaborate
specializations, but other morphological features have remained almost static.
In fleas the adults have evolved many specialized structures; the larvae have
changed little. In plants such as the Umbelliferae the flower structure has
changed little but the seeds have evolved many specializations. In mammal
lineages there seems to be a greater tendency for different parts to evolve at
comparable rates, but here again the differential rate can often be demonstrated
if highly modified features like the hominid cranium and hands are compared
with internal features such as the lungs and heart. The human heart, for example, is so similar to the dog heart that the latter is used as a test organ for
surgical innovations planned for ultimate human application.

10. The same character may evolve at different rates in different lineages.
For example, the feet of bears are primitive whereas those of horses are remarkably specialized. Both bears and horses have a common if distant ancestry. It is apparent that the feet have evolved much faster in horses than in

bears. The same is true in comparing the individual flowers of the buttercups and orchids. In the buttercups the flowers are almost uniformly a simple and primitive type; in the orchids they have evolved into a wide array of specialized types. Yet the early lineages leading to both started out as a somewhat buttercup-like organ. It is obvious that the flower has evolved extremely slowly in the buttercup line but rapidly in the orchid line. An insect example involves the stoneflies and bugs. In the stoneflies, the mouthparts are extremely primitive; in the bugs, they evolved into a highly specialized piercing-sucking organ. After the divergence of the two groups from their common ancestor, the stonefly mouthparts changed little and the bug mouthparts changed significantly at a much faster rate.

11. In regard to their overall characteristics, different lineages may evolve at different rates. An almost perfect example is found in the flowering plants. In the living magnolias almost every structure is primitive, whereas in the composites many structures are highly evolved. On the basis of fossil evidence, the magnolias have changed little from the form ancestral to all the flowering plants, but the composites have changed a great deal in many characters. In relation to the ancestral form of the mammals, the insectivores have changed little, but the horses and the whales have changed a great deal in many characters. A tabulation of many characters shows that the silver fish are likewise extremely primitive insects compared with the flies.

A corollary of premises 9, 10, and 11 is that amount of character difference cannot automatically be equated with amount of difference in geologic time. It may be in some instances but this would be a circumstance to be demonstrated, not assumed.

COMMON MISCONCEPTIONS

Several misconceptions are perpetuated concerning phylogeny, of which three especially need to be clarified.

1. *Definition of a phylogeny.* Many persons insist that a family tree must be correlated and charted against geologic time in order to qualify for the term phylogeny. Insistence on such a requirement is incorrect. Haekel, who first developed the idea of phylogeny, used his phylogenies simply to depict the supposed course of evolution and to show genealogical relationships of the organisms. It is highly desirable to make every attempt to add the time factor to a family tree, but before this can be done the family tree or phylogeny must first be constructed.

2. *Phylogeny without fossils.* There is a widespread belief that a family tree is not a bona fide phylogeny unless it includes fossils. The proponents of this

idea often have the notion that a few fossils will contain information which will, as if by magic, clear up all doubtful points in a phylogeny based only on living species. It is true that fossils often do elucidate many phylogenetic problems but, as anyone who has worked with fossils well knows, such clarification is far from automatic. Fossils do not occur *in situ* with identification labels attached. What they are and how they fit into the phylogeny must be determined by comparing their characters with those of other living and fossil relatives. Sometimes the fossils have few characters well preserved and can be assigned phylogenetically only dubiously or tentatively. Under these latter circumstances the basic phylogeny rests primarily on the living species.

3. *Two trees.* In groups having two or more distinctive metamorphic types such as larvae and adults, one investigator will arrive at one family tree on the basis of the larvae, another will construct a different one on the basis of the adults. Many feel that such a situation negates the possibility of ever knowing which tree is right. There is really no problem. Both stages are contributing information about the phylogeny, but neither is telling the whole story. Information from all stages should be combined to form a united character base for inferring a single phylogeny. The united characters would be the holomorphology of Hennig.

WORKING CONCEPTS OF PHYLOGENY

It is assumed at this point that (1) we have decided on a particular group to be investigated phylogenetically, and (2) we have established with various probabilities of correctness the identity and diagnostic characters of the available species in the group under study, whether it be a genus, family, order, or other taxon. These species should be examined according to the following operational considerations.

The Units of Phylogeny

In phylogeny the basic unit is the species, defined as a time-transect of an independently evolving phylogenetic line. Subdivisions of species may be of importance in many ways (discussed more fully in Chapter 9), but the species as a composite of the total of its subspecies or variant populations is the only logical unit in phylogeny. In this sense we deal with the species not as one or another isolated or variant population, but as the result of evolution from a preceding ancestor. Whether represented by living or dead material, the species (including its subdivisions) is the only taxon subject to experimental or inferential testing. The former would include crossbreeding trials; the latter, statistical or other analyses of populations as a measure of potential crossbreeding.

All higher taxa—genera, families, orders, and so forth—are subjective groupings of species that may or may not be monophyletic. If a genus or other taxon contains only one species, this in itself does not preclude the possibility that phylogenetically it belongs to some other taxon. When phylogenies are based on genera or higher taxa, they apply only to the individual species that were studied. Frequently it is impractical to obtain all the known species in these higher taxa. This should be no deterrent to inferring the phylogeny of the available species but it must be kept in mind that the results apply only to these species, not to the higher taxa as a whole.

Number of Species to be Treated

From an ideal standpoint, the entire world fauna or flora of a group is the only complete natural unit. But obtaining representatives of the world species is often impractical to achieve and it is always impossible to know when we might have done so. If we assemble all the *known* species of a taxon, next month or next year someone might discover a previously undescribed one. For this reason one cannot be dogmatic about insisting that a phylogeny be based on some proportion of the known species.

It is true that the closer our sample approaches the total world species, the more a resultant phylogeny can depict the entire evolutionary development within the taxon. But we must also remember that specimens of only 2 species out of 1000 known ones will give us information concerning *one* branch in the family tree. The other 998 species can be incorporated into this tree piecemeal as material becomes available. In practice, this type of situation frequently occurs. An investigator may construct a family tree based on 8 or 10 species in a particular genus. Later material of 20 more species may be discovered or become available and be incorporated into the phylogeny of the genus. If the original family tree was correct, the new material will result in simple additions to the older tree. If the original proves incorrect, information from the additional material may necessitate drastic changing of ideas and massive alterations in the tree. At this point one must remember that the end result is simply another series of hypotheses inviting still more material and/or thought prerequisite to further testing.

Characters Used

A much wider range of characters and their various states are used in phylogeny compared with studies in contemporaneous speciation. The latter are concerned with the relatively small number of characters showing differences between closely related species. In phylogeny, character states between both closely related and distantly related species come into play.

Ideally we should consider all states of all characters in the group under study, but such completeness is impossible to realize. It can be complete only for the characters we have noticed; there is moderate assurance from experience that other characters will subsequently come to our attention.

A considerable number of characters showing different states within the group can be enumerated from the literature, included in descriptions, tables, and diagnostic keys. If the group under study has not previously been investigated phylogenetically, it is almost certain that these sources will fail to provide sufficient information for adequate phylogenetic conclusions. Diagnostic treatments normally stress only enough characters to provide means of identifying the various taxa. *In phylogeny, any one character may tell a part of the story not disclosed by other characters;* hence it is necessary to use all known ones to construct a family tree having as high a degree of probability as possible. The character store of the literature should therefore be augmented by continual search for undiscovered attributes.

Ancestral and Derived Character States

If a character exhibits two or more states, one of the most critical steps in phylogeny is deciding from what ancestral condition they arose. If two states *a* and *b* are involved, there are three possible answers: *a* is the ancestral state, *b* is, or some other state was ancestral and *a* and *b* diverged from it. If we knew which was the ancestral state and which the derived, we would immediately know the direction in which the evolution proceeded with regard to the character under study. If we don't have some notion concerning this question for a particular character, we have no idea as to the direction of its evolution.

This brings us face-to-face with the question: How can we obtain evidence as to which is the ancestral or the derived state of a character? Two general considerations have been applied to this question, both concerning the progression of life.

1. Anatomically the forms of life may be arranged in series from the simplest unicells to more complex forms. One series goes from unicells to the most complex plants, another from unicells to the most complex animals. Which way did evolution proceed, from the simple to the complex, or *vice versa*? Examining other phenomena on the earth, there is nothing as yet discovered from which the higher plants and animals could have evolved except the plants and animals lower in the series. However, chemical compounds have been synthesized in presumably prebiological atmospheres that have a high probability of being the precursors of simple unicell systems (Kenyon and Steinman, 1969). Hence it is highly probable that the simple unicells evolved first, and that the more complex animals and plants represent evolutionary specializations of these simple organisms.

2. The fossil record of life is admittedly far from complete, but is developed to a sufficient extent that significant generalizations are well documented.

When this record is examined critically, it supports the preceding viewpoint. The oldest known fossils, dated as 3.1 billion years old, are remarkably like rod-like bacteria known today (Kenyon and Steinman, 1969). Other fossils up to 1.8 billion years old are likewise simple unicellular types. As we follow the fossil record upward to the present, we note a gradually increasing complexity from these earlier times to the present. The highly complex composite plants and placental mammals are recorded in relatively recent fossil beds of 150 million years or younger.

These findings support the conclusion that life started as simple forms and the more complex animals and plants evolved step by step through time. From this circumstance it is tempting to formulate the hypothesis that in any group the simplest character states are primitive and the more complicated ones are derived. But such a hypothesis is soon seen to be false. Combining fossil and living evidence, there is abundant evidence that typical flowering plants have given rise to numerous lineages in which the leaves were lost, the petals were lost, or the stamens reduced. Similarly, many typical winged insects gave rise to lineages in which legs, eyes, or wings were lost. In many vertebrate lineages the number of toes, teeth, or scales has become reduced. To compound the problem, in each of many lineages one character follows a path of reduction, another a path of elaboration.

From these considerations it is clear that the course of evolution from simple to complex is merely the statement of a trend from which there are many specific reversals, and one that cannot be assumed for any group before the phylogeny has been worked out. In judging what is the ancestral or derived character states in an unresolved group, we must therefore examine each character individually without preconceived notions or *a priori* generalizations.

Determination of Ancestral and Derived States

Lacking a specific rule for determining ancestral and derived states of a character, it is necessary to formulate workable criteria that will give the most likely answer. Three such criteria are especially useful: fossil sequences, comparisons with related groups, and group trends.

Fossil sequences. If fossil evidence in the group is extensive through a considerable span of time, the character state first appearing in the fossil record is probably the ancestral one. In dicotyledons there is a progression from either a woody to an herbaceous habit, or *vice versa.* The woody forms occur much earlier in the fossil record and presumably represent the ancestral state (Germeraad et al., 1968). Eyde (1971) cites other examples from plants, including the evolutionary trends in the secondary xylem in which the ancestral state is well documented by fossil finds.

Perhaps one of the best examples of using the fossil record to infer ancestral and derived states concerns the vertebrates. Here there is a succession from finless notochordates to the fishes with four lateral fins, to the various tetrapod types with fingered appendages, to the whales having fins. Which were ancestral, fins or feet? The fossil record indicates clearly that fish-type fins occurred before feet, and that whale-like fins occurred long after feet had evolved. The summation of the fossil record combined with detailed studies in comparative morphology indicates that fish-like fins evolved first, then tetrapod legs evolved from them, and later modified whale fins evolved from tetrapod legs.

To cite another example, in the Paleozoic ectoproct family, Aisenvergiidae, the zooecial tubes have character states ranging from the poorly organized condition found in *Lamottopora* to the highly organized condition found in *Aisenvergia* (J. P. Ross, 1967). On the basis of the character states alone, it would be impossible to decide whether evolution had been from the poorly organized state to the more highly organized one, or *vice versa*. The fossil record indicates the former with a high degree of probability (Fig. 1).

In the living land vertebrates there is an orderly array of teeth from the undifferentiated numerous teeth of the reptiles to the various specialized types in the mammals. The simple reptilian teeth occur first in the record, hence they are probably the ancestral state (Romer, 1959). A magnificent example occurs in the elephants (Fig. 2) in which first occurrences document the ancestral states of many skull characters and the states that occurred simultaneously, including the loss of certain teeth, the specialization of others, and changes in the skull. In the wings of insects, the fluted wing exemplified by mayflies and dragonflies appears first in the fossil record and is probably the ancestral state.

But if the fossil record of the group is meagre, first occurrences can be misleading. In the caddisfly family Glossosomatidae, several living species have wings that are obviously much more ancestral than those of the few fossils, all of which occurred in only a single time horizon.

Comparisons with related groups. If two or more character states occur within a group having an indecisive fossil record, it is difficult if not impossible to know from a comparison made within the group which character state is ancestral and which derived. If, for example, certain species in a group have smooth leaf edges whereas others have serrated leaf edges, there is no indication from these data alone whether *smooth* or *serrate* is the ancestral state. Comparisons within the group (often termed *in-group* comparisons) need supplementation by comparisons with character states found in other groups (*ex-group* comparisons).

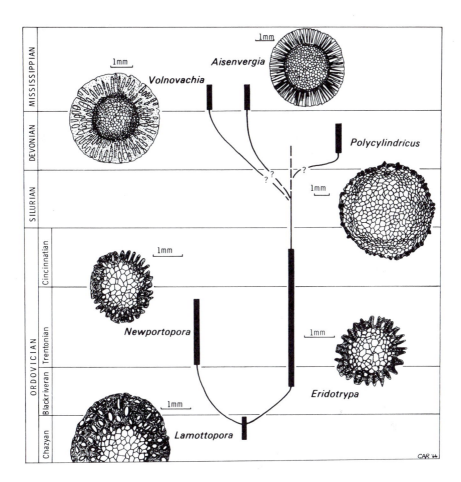

Fig. 7.1 Six Paleozoic genera of the family Aisenvergiidae (Ectoprocta) showing their stratigraphic and probable phylogenetic relationships. (From J. P. Ross, "Champlainian Ectoprocta (Bryozoa), New York State," *Journal of Paleontology*, 41:632-648, 1967. Reprinted with permission of the Society of Economic Paleontologists and Mineralogists.)

In these ex-group comparisons, it is considered that if one of two or more character states in one group occurs in other closely related groups, it is probably the ancestral one. In the leafhopper genus *Exitianus,* the plates of the male genitalia exhibit two states (Fig. 3), triangular with lateral setae (a) or rounded without lateral setae (b). In the closely related genus *Nephotettix* the plates are triangular and have lateral setae (c). The triangular, setose plate

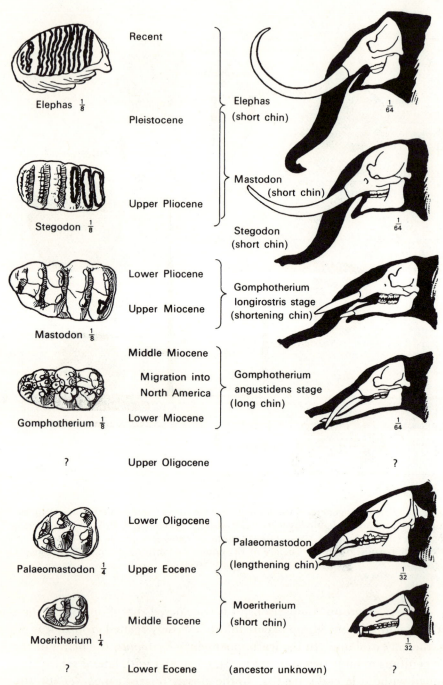

Fig. 7.2 Evolution of the Proboscidea. Right, a series of known skulls with the flesh restored in silhouette; left, last lower molar. (Adapted from Lull in Scott, *A History of Land Mammals in the Western Hemisphere.*)

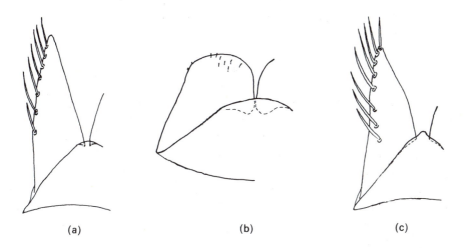

(a)   (b)   (c)

Fig. 7.3 Plates of the male genitalia of the closely related leafhopper genera *Exitianus* and *Nephotettix*. (a) *E. indicus*; (b) *E. zuluensis*; (c) *N. apicalis*. (From H. H. Ross, "The evolution and dispersal of the grassland leafhopper genus *Exitianus*, with keys to the Old World species (Cicadellidae: Hemiptera)," Bull. Br. Mus. (Nat. Hist.), Ent. 22 (1) : 1–30, 1968. Reprinted with permission of the British Museum (Natural History).

is probably the ancestral state. In this instance the probability is increased because in other both closely related and distantly related genera the plates are triangular and setose.

The reasoning behind the value placed on ex-group comparisons as indicators of ancestral *versus* derived states is based primarily on the fossil record. This record shows that various forms of life have each given rise to diverse types, and types characterized by new, derived character states. For example, the earliest mammal probably had carnivore-like teeth similar to those of its reptilian therapsid ancestor (Romer, 1959). From this stock arose lineages having such diverse dentition as have cows, horses, elephants, and baleen whales. Although meagre, the fossil record indicates that the earliest flowering plants probably had magnolia-like flowers. From these evolved extremely diverse floral structures exemplified by orchids, grasses, mints, and composites.

These models of evolution led to the inference that different lineages arising from a common ancestor diverge characterwise. If some character in the group under study has two or more character states and one of these states occurs in a related group, then species having the state common to the two groups are probably closer to the point of divergence of the groups. A corollary is that the common or shared character state is the ancestral one.

The reliability of this ex-group comparison method in determining the ancestral and derived states of a character is proportional to the investigator's knowledge of forms related to the group under study. The wider this knowledge, the greater is its application to a variety of characters. For example, hind wings of the caddisfly genus *Agapetus* exhibit a wide range of venation from many veins to few. A remarkably similar set of hind wing character states are found in several other families. In certain caddisfly families only the fuller vein complement occurs, indicating that this is the ancestral state, although there might be some doubt as to the logic of conclusions based on these in-group phenomena. If one consults data from insects in general, the same conclusion is reached. Here it was desirable to compare not only genera within the family, but the family with the orders of an entire class. If any one of these clues later proves to be incorrectly resolved, we must reexamine all conclusions based on it.

In groups whose chromosomes have localized centromeres, cytological evidence may be useful in reconstructing points of the phylogeny. Fusions, translocations, and inversions are especially useful. In using these cytogenetic criteria, however, it is essential to determine the ancestral and derived states of each contrasting set of character conditions. This must usually be done by comparisons of the states occurring in the group under study with those found in other groups.

When the different states of a character are compared, they frequently form an intergrading series of steps from one extreme to the other. The series may consist of only 3 or 4 steps, but sometimes may embrace 10 or more. If each step is distinctive for one or more species, the series can be regarded as a *phenocline*. These series have previously been termed *morphoclines* (Maslin, 1952), but they occur in biochemical phenomena also (Brand et al., 1972) and presumably in behavioral and other phenomena. Brand et al. (*op. cit.*) proposed *phenocline* as a term that would apply to all types of phenomena.

If a phenocline in the group under study has a character state common to a phenocline in another group, the common state is probably the ancestral state and the others are derived (Maslin, 1952). This is simply a special case of ex-group comparison. If two apparent phenoclines occur in the group under study, there is no way from this evidence alone to know whether there are two phenoclines or only a single one embracing the entire series of character states. Only a comparison with character states in other groups or evidence from some other characters can solve this dilemma. For example, in the leafhopper genus *Erythroneura* the styles of the male genitalia form a phenocline ranging through seven states (Fig. 4 parts (a) through (g)). From this series alone it is not possible to decide whether only one phenocline is involved, embracing all eight steps, or two shorter ones. When these states are compared with those in re-

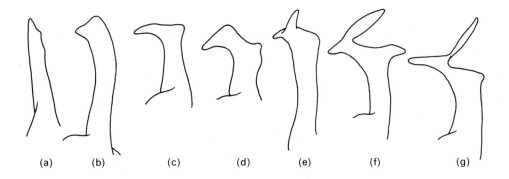

Fig. 7.4 A phenocline of the various types of style apices found in the leafhopper genus *Erythroneura.*

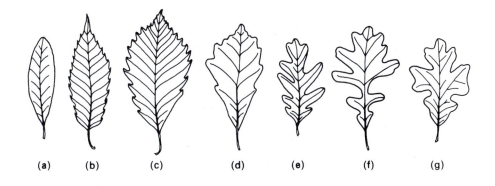

Fig. 7.5 A phenocline of the various shapes of leaves found in the white oak complex (*Lepidobalanus*) of the genus *Quercus.* (a) *virginiana,* (b) *muelenbergii,* (c) *michauxii,* (d) *bicolor,* (e) *alba,* (f) *macrocarpa,* (g) *stellata.*

lated genera, it is apparent that state (c) is ancestral, hence there are two phenoclines, one from (c) to (a), the other from (c) to (g).

A comparable phenocline occurs in the leaf shape of the North American white oak complex *Lepidobalanus* of the genus *Quercus* (Fig. 5), progressing from the smooth-margined elliptic leaf of *Q. marilandica* (a), to the serrate leaf of *Q. muehlenbergii* (b), and by small steps to the broad, lobate leaf of *Q. stellata* (g). When this phenocline is compared with genera in other members of the order Fagales, it is seen that character state (b) is similar to the leaves of *Fagus* and *Castanea,* belonging to the same family Fagaceae as does *Quercus.* Furthermore, of all members of the phenocline, (b) most closely resembles the leaves of *Alnus, Betula, Carpinus* and *Ostrya* of the related family Betulaceae. From this ex-group comparison, it seems evident that in Fig. 5 the leaf shape (b) is the ancestral state for the complex *Lepidobalanus.*

Unusual difficulty may be encountered in determining the ancestral and derived states of characters arising within a group. In such circumstances there is no possibility of making comparisons with other groups. Frequently, however, other characters will indicate which species are lowest on the family tree and this may indicate which is the ancestral state of the *de novo* characters.

Group trends. In various groups of organisms, characters appear to follow the same evolutionary development in many independent lineages. Thus in insects, the loss of veins in the wings appears to occur almost automatically with decrease in absolute body size. A minute beetle or wasp or moth will almost invariably have a great reduction in wing venation compared with larger members of the same group. Similar trends are commonplace in the flowering plants, notably concerning structures of the flower itself. Thorne (1963) has listed many of these trends, which resolve into truisms that, in flowers, homologous parts such as petals and sepals tend to fuse rather than divide and that stamens tend to decrease in number. Other noted trends are equally illuminating.

On logical grounds, trends of this type are useful as a start in orienting our first thoughts as to what are the ancestral and derived states of any particular character, but they cannot be accepted as *laws* because we have no assurance that sometimes they might have progressed in reverse, as Eyde (1971) has pointed out in plant examples. Trends are therefore useful but require additional evidence from other sources before they can be considered applicable to a particular group. In insects, for example, wing veins may be reduced, but in subsequent lineages they may be increased, as has happened in the lacewing flies or Neuroptera.

Terminology

The examples given above serve as an excellent background for discussing terminologies commonly applied to character states. They are often differentiated as *primitive* vs. *specialized,* but this frequently conflicts with another categorization of *simple* vs. *specialized.* In the latter context, *specialized* implies an elaboration of a character, whereas when opposed to *primitive* it may be the loss or simplification of a character. As a result, when either of these two systems are used as a basis of definition, there is bound to develop confusion over the meaning of *specialized.* The result would be semantiological arguments in which neither party understood what the other meant.

It is therefore much simpler to use the opposites *ancestral* and *derived,* because neither term denotes any particular direction of change, only that some sort of character change did occur. Recognizing this problem, Hennig (1950) coined the term *plesiomorphic* for the ancestral state and *apomorphic* for the derived state. I personally prefer the terms *ancestral* and *derived* because they continually remind one of the evolutionary relationships of the two terms. This relationship of the terms to character states may change, as is immediately seen with regard to phenoclines. In succeeding steps of a phenocline, the derived state of one step becomes the ancestral step of the next, and so on. I believe the situation is handled better with common well-understood adjectives rather than with a specialized terminology.

The results of applying the previous concepts to a particular group under study should be a summary or tabulation of (1) the ancestral and derived states of certain individual characters and (2) the failure to achieve this distinction with any degree of probability for other characters. In practice, it is possible to ascertain the ancestral states of most characters with a high degree of probability. The most difficult situations are those involving characters that arose *de novo* in a group, in which case comparisons of character states in other groups are impossible.

The next problem is to use this information in reaching inferences concerning the probable course of evolution, that is, the phylogeny, of the species units of the group.

PHYLOGENETIC INFERENCE

Ideally, phylogeny should be inferred by comparing all character differences and similarities between all the species, past and present, in the biological universe. It is impossible to do this for all species simultaneously because of several reasons. In the first place, we don't have information of comparable (homologous) character states throughout the entire biological spectrum. Bacteria, for example, have well-explored biochemical characteristics, but have neither flowers, legs, nor wings. Plants, mammals, and insects, on the other hand, have well-explored structures such as flowers, legs, or wings, but their biochemical characteristics in general have formerly been poorly known. As a matter of fact, many higher plants and animals (and some bacteria) are known only from dead preserved specimens or from fossils, from which the biochemical characters cannot be derived.

In the second place, we have as yet no techniques for handling such a mass of information in a fashion that will allow clear visualization of all character changes from ancestral to derived. The best practical solution is to organize the species into groups small enough that all their known pertinent characters can be visualized clearly. It would of course be helpful if such a grouping produced clusters of species that were closely related and inclusive monophyletic units; that is, species groups which (1) all descended from a common ancestor, and (2) included *all* the known descendants of the common ancestor. This aim can best be approximated, but not logically demonstrated, by four methods.

1. *Species keyed out together in some existing key.* This method is highly erratic. In general, botanists try to make their keys express what is known of the phylogeny, in which case the method would be relatively effective. On the other hand, in their keys, zoologists generally stress clarity of diagnosis, in which differences rather than similarities are stressed. In instances in which convergences were frequent, such keys would give a poor picture of monophyletic groupings.

2. *Adansonian methods.* These include all those techniques described as *numerical taxonomy* in which all recorded characters are expressed as a statistic of cluster analysis or other criteria. When the ancestral and derived character states have been established for few or no characters, as in the bacteria and fungi, this is undoubtedly the best preliminary sorting method. The rationale for this method is that similarity in many characters would seem to indicate similarity in genetic composition, and this in turn could indicate a monophyletic origin of species clusters.

In some of the early studies, for example, that of Michener and Sokal (1957) on a group of bees, the numerical results paralleled and, in spots, improved upon existing phylogenies. In some later studies, for example, Rohlf's (1963) analysis of mosquitoes, the use of characters from different life history stages gave divergent results. In their review, Sokal and Sneath (1963) stressed the use of numerical taxonomy as an aid in classification.

Since the appearance of the work by Sokal and Sneath in 1963, many techniques of numerical taxonomy have been explored by a large number of investigators (e.g. Rohlf, 1967; Moss, 1967, 1968). More recent reviews are by Jardine and Sibson (1971) and Blackith and Reyment (1971). Moss pointed out that various computer programs did not produce the same results from the same data; for example, that certain species clustered together on the basis of one program would be separated widely in the phenogram on the basis of some other program. In his mite studies (1967) he found one of the three-dimensional models the most accurate in placing units according to character similarity (Fig. 6). There is thus a need for further methodological research in this field. Heiser et al. (1965) found that grouping of species in the plant genus *Solanum* by numerical methods disagreed seriously with groups suggested by fertility tests for certain species. They concluded, however, that the numer-

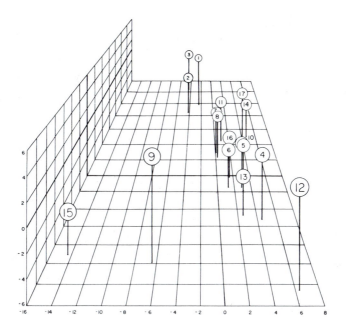

Fig. 7.6 Grouping of 17 species (OTU's) of mites by numerical taxonomic methods using Rohlf's method of three-dimensional centroid factor projection model. (From W. W. Moss, 1967.)

ical methods provided useful information in correctly grouping hybrids and autopolyploids.

3. *Possession of a single derived character state.* If the derived character state is an additive one or an elaboration of one, the grouping has an excellent chance of being monophyletic. Here the rationale is that *new* characters arise infrequently and *usually* not in parallel. The *usually* is emphasized because in a goodly number of instances a character state has evolved twice and therefore might do so in various unsuspected circumstances. If the derived character state is a loss, it may offer little evidence that the grouping of species represents a monophyletic unit. Parallel losses are too frequent to be considered decisive evidence of monophyly.

4. *Possession of several striking and presumed derived characters.* Theoretically this would be the ideal preliminary grouping of species for further phylogenetic analysis. As will be explained later, such a grouping could involve some simple errors not detectable from an examination of a collection of spe-

cies. But as a starting point it has the best chance of resolving species into monophyletic units. The greater the number of correlated derived characters, the greater the probability that the grouped species are a monophyletic unit.

Each group of species resolved to the best of our ability according to the above precepts becomes the basis for further phylogenetic analysis.

ANALYTIC PROCEDURES FOR CONSTRUCTING FAMILY TREE

At this point we assume two items:

1. The species under study have been grouped into quasi-phylogenetic units according to one of the procedures set out above.

2. From previous studies outlined in Chapter 6, we know which species are

   a) Apomicts

   b) Polyploids of different ploidy levels

   c) Bisexual species of the lowest ploidy level. These would be isoploids. The greatest number are diploids, but the lowest known level could be higher. For example, all living bisexual species of *Equisetum* have a chromosome number of 108, thought to be decaploid or higher, yet this is the lowest known ploidy level in these living plants.

Because the lower isoploid levels are the overall continuing lineages from which higher bisexual ploidy levels and various types of apomicts have arisen, it is simplest to work out the phylogeny of the lower levels, then superimpose the higher ploids and the apomicts on the resultant tree. In the situations depicted in Fig. 7, for example, it is easiest to first infer the phylogeny of the diploid species in (a) and the bisexual species in (b), then to add the polyploid species or the apomict species to the respective phylogeny.

Bisexual Lower-Level Isoploid Species

The phylogeny is inferred by comparing combinations of ancestral and derived character states in the species under study. Characters used must satisfy two conditions: (1) each must exhibit two or more character states (otherwise they would contribute no intra-group information) and (2) the ancestral state of each character must have been inferred with some degree of probability.

If living material is available for study, it affords a much greater wealth of study characters than does dead material. Especially useful are cytogenetic, behavioral, and biochemical characters. Cytogenetic characters exhibit certain peculiarities of analysis, and are treated below as a special section. Biochemical and behavioral characters are phenotypic expressions of the genome, and in a logical sense are treated like any other phenotypic characters.

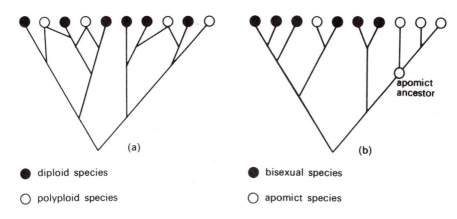

Fig. 7.7 Hypothetical family trees involving diploid and polyploid species in (a), bisexual and apomict species in (b).

If only nonliving material of most of the species is available for study, then only those characters detectable in dead specimens can be employed for the group analysis as a whole. This does not preclude the possibility of using live characters for inferences concerning small complexes within the group for which live material is available.

Cytogenetic Characters

In this category, two phenomena have proven to be unusually valuable for inferring phylogeny. These are centric chromosome fusions and inversions. Translocations theoretically should be also, but they occur infrequently, at least in wild species of *Drosophila* (Stone, 1962).

Centric Fusions

These are usually considered to be the derived state; free chromosomes, the ancestral one. There is sufficient doubt on this point, however, that in studying any particular group care must be taken to establish the ancestral state by comparisons with related groups (White et al., 1964).

Inversions

Inversions are considered to be unique events that are nonreversible. To date, no two identical inversions have been discovered. Further, when the complexity of the DNA strand is realized, the possibility that two inversions would arise each embracing exactly the same section of the chromosome is considered so remote as to be improbable to the highest degree. However, two such identical events might eventually be found.

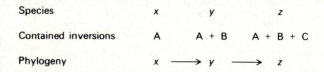

Fig. 7.8 The utilization of additive inversions in deducing phylogeny.

If the inversions are nonoverlapping, nonhomologous, and each or a discrete group are confined to a single species, they offer no evidence of phylogenetic grouping. If, for example, inversion A occurred only in species *x*, inversion B only in species *y*, and inversion C only in species *z*, the inversions would provide no information about the phylogeny of these three species. On the other hand, if the inversions occurred in a cumulative series in the three species, the phylogeny would be apparent if the ancestral state were known. Thus if inversion A occured in species *x*, A and B in *y*, and A, B, and C in *z*, and if the ancestral state were inversion A only, the order of origin would be from *x* to *y*, and *y* to *z* (Fig. 8).

If the inversions are overlapping, the order of origin may be immediately apparent. For example, if three species have the following arrangement of loci along a homologous chromosome:

species *x* — ABCDEFGH
species *y* — ADCBEFGH
species *z* — ADCFEBGH

and we have assurance that the *arrangement in species x is ancestral,* it is obvious that first the region BCD inverted and became established in species *y*, that subsequently species *z* arose from species *y*, and in it the BEF arrangement of species *y* inverted, thus:

ancestral — ABCDEFGH
first inversion — ADCBEFGH
second inversion — ADCFEBGH

where the inverted regions are designated with braces.

If only the ancestral state and that in species *z* are known, the past or present occurrence of species *y* can be predicted.

As the numbers of overlapping inversions increase, the difficulties of unscrambling their past history also increase. The method described above is therefore of limited application, but has nevertheless proven to be an extremely useful tool.

Application of the above cytogenetic criteria to adducing phylogeny was first explained and used by Hsu (1952) and further diagrammed by Patterson and Stone (1952). Using overlapping and nonoverlapping inversions and centric fusions, Hsu worked out the phylogeny of the nine species of the *Drosophila virilis* group and predicted the past existence of three hypothetical forms necessary to explain the chromosomal arrangements found in the known living species. Present day species corresponding to these hypothetical ancestors (the "primitives" of Hsu) are not yet known, but they may eventually turn up.

The reasoning of Hsu in these phylogenetic studies emphasized the need for sufficient knowledge of some character states outside the group in order to get a probable answer as to which was the ancestral and which the derived state of the cytogenetic characters.

Phenotypic Characters

Whatever the type, these lack the high probability of uniqueness characterizing chromosomal rearrangements. As Dobzhansky (1970) pointed out, mutants arising in different species frequently produce similar phenotypic changes. But these mutants need not be homologous because phenotypically similar mutants are often produced at different loci in the same species. This is especially true of polygenic characters in which the mutant types involve losses. For example, if a protein C is transcribed by a certain extensive portion of the DNA sequence, a mutant at one locus might disrupt the protein synthesis, and would be a new allele, c_1. In another population or another species, a mutation at another locus might disrupt the synthesis, constituting a nonhomologous allele, c_2, that was phenotypically similar to c_1. Similar types of mutations at different loci could give rise to the disruptive alleles $c_3, c_4, \ldots, c_n$, all nonhomologous yet all producing identical phenotypes. Recurrent mutations of this type are relatively common (Solbrig, 1970). These circumstances do not mean that loss characters are of no value in inferring evolution, but that they may be misleading.

When a new character state presumed to be polygenic becomes completely established in a lineage, this would be the result of a series of genetic events under the influence of natural selection in a particular setting. There is a considerable probability that such a series of events would not occur twice, in other words, that the new character state approached the nature of a unique event. The more complex the situation, the more probable that it would be unique. When many character states have coevolved to form a complex organ or system, the probability of this being unique is extremely high. Examples are the wings of insects with their attendant muscles and body subdivisions, the xylem-cambium-phloem system of the higher plants, and the backbone of the vertebrates.

With these limitations and possibilities in mind, phenotypic character states of different species may be compared and the phylogeny of the species inferred. Because character states vary in the probability that they represent unique events, it is necessary to use as much evidence as can be brought to bear on any particular group under study.

Hypothetical Ancestors

These are invoked in the same fashion as outlined above for cytogenetic characters to express a combination of character states needed to explain an inferred phylogeny. To illustrate their use, let us assume that we have two species x and y, that sorted out together and possibly arose from a common ancestor A (Fig. 9a). Suppose that species x has ten teeth, y has 8, and that 10 is the ancestral state. When this base information is plotted on Fig. 9a, using a for ancestral and d for derived (Fig. 9b), it is indicated that the ancestor of the two species possessed the ancestral state a, that species x did not change from the ancestor, but species y did. Adding character 2, number of claws, with two states, 5 claws (ancestral or a) and 4 claws (derived or d), we obtain one of two results. If species x has 5 claws (a) and species y has 4 claws (d), the picture of the relation of the ancestor to its daughter lineages has not changed (Fig. 9c). In these characters the ancestor is like one of its daughter species.

If on the other hand species x has four claws (d) and species y has five claws (a), we obtain an entirely different solution (Fig. 9d). This indicates that the ancestor had the ancestral condition for both characters and was therefore *different from either daughter* species. The ancestor could have been either heterozygous or homozygous for either trait. Here I am following the logic of Stone (1962). If character 1 is ancestral in species x, heterozygous for both ancestral and derived in species y, Fig. 10a, then the simplest solution is that the ancestor had the ancestral condition and the derived condition arose in the lineage leading to species y. If both species x and y are heterozygous for the ancestral and derived states, Fig. 10b, then the ancestor was also heterozygous. If the species x has only the derived state and species y is heterozygous for ancestral and derived, Fig. 10c, then the ancestor was undoubtedly heterozygous also; the line leading to species x had become homozygous for the derived condition. Applying this to Fig. 9d, in the absence of evidence indicating heterozygosity in species x and y for either character, the ancestor is considered homozygous for the two ancestral states. If evidence indicates that a succession of ancestors were heterozygous for different states of the same character, as Throckmorton (1962) inferred for several characters in *Drosophila,* the evolutionary consequences are so uncertain that the character cannot be used for inferring phylogeny; its evolutionary development can be deci-

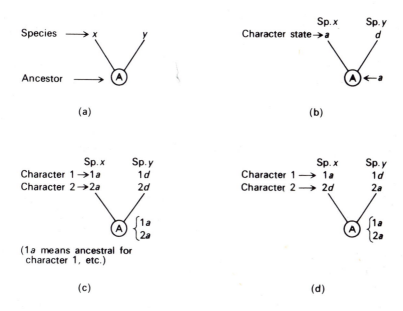

Fig. 7.9 Inferring the hypothetical ancestor of two species. For explanation, see text.

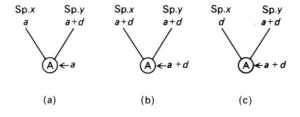

Fig. 7.10 Hypothetical ancestors involving heterozygous ancestral (*a*) and derived (*d*) character states.

phered only after the group phylogeny has been inferred on the basis of other characters.

Hypothetical ancestors, primitives, or whatever else they may be called, represent a peculiar dual concept. First, each one can be defined only in terms of those characters for which the state or condition in the particular ancestor has been inferred. If this has been done for only ten characters, the ancestor can be defined only in terms of those ten. In this fashion each hypothetical ancestor represents a *character pool,* comparable to a gene pool, from which subsequent daughter lineages or species evolved. Second, such a group of character states could not have existed in a vacuum; they must have been carried by viable organisms constituting a past population or species. Hence these ancestors were past biological entities that can be defined only in terms of their inferred character pool.

The hypothetical ancestors are our only means of inferring certain probable evolutionary stages in lineages when such stages are not known as living or fossil species. They are mental constructs that serve as "stand-ins" for sequential species and hence give a prediction as to what probably occurred in the past history of a group. Predictions arrived at in this manner may then become subject to the test of subsequent observation. Thus Hamilton (1971) postulated certain character states for the progenitors of the insect family Cicadellidae based solely on Recent species; later, a Cretaceous fossil was found that had these character states, representing a new taxon, the family Jascopidae.

Sequential species and hypothetical ancestors share an important circumstance. We can infer differences between successive stages in a lineage, whether these stages are represented by living species, fossil species, or hypothetical ancestors, but we have no means of knowing when in the time interval between stages the differences occurred or how rapidly or in what order they evolved. For example, in part of the *Drosophila montana* complex, Stone (1962) inferred that in ancestor III the second chromosome lacked inversions G, H, J, and K, and that in ancestor IV these inversions were present in a homozygous state (Fig. 11). Without additional information, it is impossible to know in which order the four inversions arose, and when in the time journey between ancestors III and IV the various events happened.

Methods for Constructing the Phylogeny

This is done by first listing the various character states under each species, then making inferences on the basis of character state distributions. It is important that one use only characters for which the ancestral and derived states are known. States of the other characters can often be inferred after a highly probable phylogeny has been constructed (see Chapter 12). It is also necessary that each character used exhibit two or more states within the group. If the character is uniform throughout the species, it offers no help in deciphering phylogeny within the group.

The states ancestral and derived are to be considered relative within each group. For example, if in a group of several insect species the antennae are 9-segmented in some and 6-segmented in others, a comparison with other

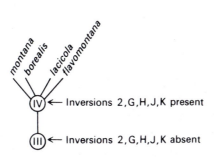

Fig. 7.11 Inversions in two hypothetical ancestors (primitives) III and IV in the *Drosophila virilis* group. (Redrawn from Stone, 1962).

groups will indicate that the more primitive insects have up to 30- or 40-segmented antennae. Hence, *within the study group*, the 9-segmented antenna is probably ancestral. Both the 9-segmented and 6-segmented antennae are derived states compared with outside groups, but within the group the higher number of segments represents a state closer to the primitive and hence is the ancestral condition as regards the study group.

In the interests of clarity, the following examples of situations encountered begin with simple ones involving study groups of several species, then proceed to inter-group problems. Many of these examples are similar to those given by Hennig (1950,1966). The general aim is that explained by Ross (1937), to arrange the taxa in a family tree such that derived character states arise from ancestral ones, and not *vice versa*. This is done by grouping together species sharing a particular derived state on the assumption that they arose from a common ancestor possessing that state, then grouping together other species that may have a common derived state. As these groups are joined together by the following practices, there should result a gradual progression of groups having more and more derived states arising from the group ancestor in which all states are ancestral.

Combinations of Three Species and with Only One Character Difference

In a system of three species, x, y, and z, a character exhibiting two states, ancestral (a) and derived (d), can have only two combinations:

$$2 \text{ species} - d, \ 1 \text{ species} - a, \tag{1}$$

or

$$1 \text{ species} - d, \ 2 \text{ species} - a. \tag{2}$$

The first combination is exemplified by the following distribution of any character 1:

sp.x	sp.y	sp.z
ld	ld	la

From this it is seen that sp.x and y share a derived character and both arose from an ancestor having this derived state (Fig. 12a). When this is drafted, we can compare the ancestor of x and y with species z and determine that the ancestor of the group must have had the ancestral state of the character and in this particular was like species z.

The second combination is exemplified by the following distribution of any character 1:

sp.x	sp.y	sp.z
ld	la	la

It is tempting to resolve this distribution in the same manner as in Fig. 12a, grouping sp.y and z as the products of an ancestor having the ancestral state of the character, as depicted in Fig. 12b, but this runs into a logical fallacy. The derived state ld could have arisen from any point on the tree with state la. We must therefore depict species x as having arisen from any one of three places in the tree, (1) the group ancestor, (2) any point on the lineage leading to species y, or (3) any point on the line leading to species z. We have no evidence that will group species y and z into a common branch above the group ancestor. If we did so, we would find that the ancestor of that fork was exactly like the group ancestor and that we had not changed the tree at all; two of the possible origins of species x would still be above the point of dichotomy. The only honest solution is Fig. 12c, which can be expressed more simply as Fig. 12d.

Two conclusions emerge from this treatment:

1. the second distribution of two character states a and d of only one character cannot solve the phylogeny of three species, and

2. species cannot be grouped into monophyletic units above the group ancestor on the basis of shared ancestral character states.

Hennig (1950) was the first systematist to make this important distinction. Failure to observe it is probably one of the commonest mistakes made in phylogenetic studies.

Combinations of Three Species and Two Characters

The dilemma of the second character distribution and of Fig. 12c can be solved only by adding information from other characters. Adding character 2 having

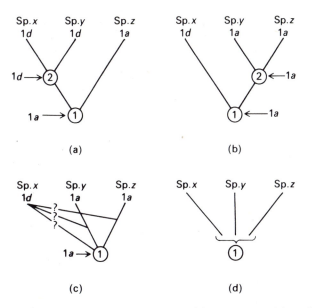

Fig. 7.12 Phylogenetic groupings based on derived (a) and ancestral (b,c,d) character states.

two states a (ancestral) and d (derived) will result in two types of distribution:

$2d$ only in any two species, x and y, or x and z, or y and z (1)

$2d$ only in one species, x, y, or z.   (2)

When combination (1), using $2d$ in x and y, is added to Fig. 12b, we get Fig. 13a. Now that x and y share a derived $2d$, it is seen that x and y are a monophyletic unit and that the derived state of character 1 arose from the lineage leading to x, arising from an ancestor having the character combination $1a + 2d$. If in this combination we substitute $2d$ in x and z, then x will be linked with z rather than y necessitating a juggling of the species in the family tree. In each case, one species will have the same character states as the group ancestor.

When the combination $2d$ in y and z is added to Fig. 12b, we get Fig. 13b. Here y and z can be grouped as a monophyletic unit possessing $2d$, arising from common ancestor 2 possessing $1a + 2d$. When this y-z ancestor is compared with x, it is seen that the group ancestor possessed the combination $1a + 2a$. As in Fig. 9d, so here also the group ancestor possessed a combination of character states different from that in any known species in the system.

When combination (2) is added to Fig. 12b, we encounter some peculiar situations, and it is necessary to consider all three possibilities separately.

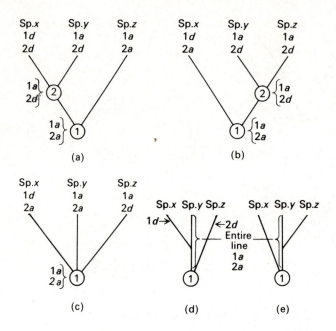

Fig. 7.13 Phylogenetic solutions involving three species and two characters.

1. 2*d* only in *x* results in the distribution:

	x	*y*	*z*
Character 1	1*d*	1*a*	1*a*
Character 2	2*d*	2*a*	2*a*

It is immediately apparent that this combination is as unresolvable as Fig. 12b and for the same reasons. Information must be sought from other characters.

2. 2*d* only in *y* or *z* give the same result shown in Fig. 13c, using 2*d* in *z*. The group ancestor is 1*a* + 2*a*, the same as *y*. It is obvious that *x* arose from a 1*a* + 2*a* form in which character 1 became derived and character 2 did not, and that *z* also arose from a 1*a* + 2*a* form in which only character 2 became derived. On the basis of these characters alone, *y* continued as a 1*a* + 2*a* lineage similar to the group ancestor. We have no way of determining from these data in which order the lineages of *x* and *z* arose. All three lineages could have diverged at the same time, as indicated in Fig. 13c, or lineage *z* could have diverged before lineage *x,* as in Fig. 13d, or lineage *x* could have diverged before lineage *z,* as in Fig. 13e. The line from ancestor 1 to species *y* remained unchanged in the two characters considered. The evidence from these two characters alone presents no logical choice between these three alternatives.

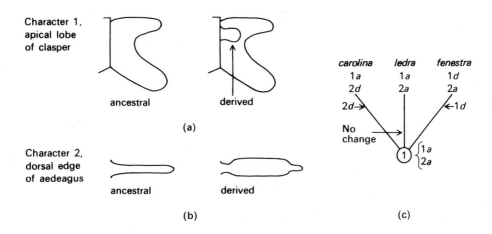

Character 1, apical lobe of clasper — ancestral / derived (a)

Character 2, dorsal edge of aedeagus — ancestral / derived (b)

carolina   ledra   fenestra
1a   1a   1d
2d   2a   2a
(c)

Fig. 7.14 Phylogenetic solution of a portion of the caddisfly *Rhyacophila carolina* complex.

It is surprising how often this situation arises. For example, in the three species comprising the *Rhyacophila carolina* complex of Nearctic caddisflies, two differences have been observed in each of two characters. The apical segment of the male clasper, character 1 (Fig. 14a), either has an inner lobe at its base or has none. Comparing these conditions with those in other species groups in *Rhyacophila* indicates that the nonlobed condition is the ancestral state, the lobate condition the derived. The upper edge of the aedeagus, character 2 (Fig. 14b), is either narrow or flared into a flat top. Again, a comparison of these two conditions with those in related groups indicates that the narrow condition is the ancestral state, the flared top the derived state. The distribution of these character states for the three species in the complex (Fig. 14c) shows that *R. fenestra, ledra,* and *carolina* have exactly the same relationships as *x, y,* and *z* in Figs. 13c, d, and e.

The dilemma of Figs. 13ç and 14c can be resolved by only a limited number of considerations: (a) certain state distributions of a third character, (b) amounts of difference in character states, and (c) certain paleontological circumstances.

a) A third character having an ancestral and a derived state may resolve the dilemma if the distribution is different from that of either character 1 or character 2. Of the possible combinations, only two satisfy this requirement:

derived state (*d*) in any two species,   (1)

derived state (*d*) only in species *y*.   (2)

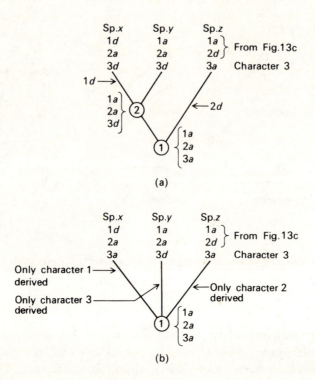

Fig. 7.15 Phylogenetic solutions involving three species and three characters.

Combination (1) in which character 3 is derived (3*d*) in species *x* and *y* produces the phylogeny shown in Fig. 15a, resulting in two successive dichotomies in which both hypothetical ancestors are unlike each other or any of the three daughter species *x*, *y*, and *z*. If *y* and *z* have 3*d*, the phylogeny is the mirror image of Fig. 15a. The other permutation of combination 1, having 3*d* in *x* and *z*, produces two successive dichotomies in which *x* and *z* arise from a parent having 3*d*, and *y* is ancestral in all three characters and similar to the group ancestor.

Combination (2) results in the trichotomy shown in Fig. 15b, in which each of the species *x*, *y*, and *z*, has evolved a derived state in a different character, all arising from the same common ancestor which is different from any of the three daughter species.

The combinations of the states of character 3 that do not differ from those of characters 1 and 2 produce exactly the same result as Fig. 13c, lead-

ing to the necessity for seeking yet other characters to resolve the dilemma.
b) If the difference between the ancestral and derived states of character 1
is much greater than that between the ancestral and derived states of character 2, there is some reason for considering Fig. 13e as the correct solution. If
the reverse were true, then Fig. 13d would be favored. The assumption invoked is that amount of difference in character evolution is probably equivalent to amount of geologic time. In his studies on rates of evolution, Simpson
(1944) presented excellent evidence that these rates are different in various
evolving lineages. Tabulating probable evolutionary rates for many lineages,
Simpson found that most lineages evolved at moderate rates, a few at unusually low rates, and a few at unusually high rates. These categories of tempo he
designated as *horotely* (average rate), *bradytely* (unusually slow rates), and
tachytely (unusually fast rates). Because of the relatively infrequent occurrence of unusually slow or fast rates, it is probable that amount of difference
does actually give a measure of geologic time, *if no other circumstances con-
travene this conclusion.* In other words, if no other information gives clues as
to the relative time of origin of various phylogenetic branchings, then the degree of character difference between lineages arising from the parental stem
gives a clue as to the probable (not logical) sequence of branching.

To prevent others from reaching misleading conclusions, when an investigator uses amount of difference to choose between Figs. 13c, d, and e, he
should make the reasons for his choice perfectly clear. Such a statement will
alert everyone as to the possibility of alternative arrangements.

c) If either species x or z occurred earlier in the fossil record, it is an indication that it branched from the lineage of y earlier than the other. This circumstance is not a probability unless the fossil record for the group is unusually
good and the time span involved is considerable. The problems encountered
are summarized in Chapter 8.

Conflict of characters. Returning to character state permutations involving
three species and two characters, each with two states, one type of combination leads to conflicting conclusions. Let us consider Fig. 12a. If we add to
this combination a second character with the derived state in y and z, we get

	x	y	z
Character 1	d	d	a
Character 2	a	d	d

Character 1 indicates that species x and y arose from a common ancestor as opposed to z (Fig. 16a), whereas character 2 indicates that y and z arose from a
common ancestor as opposed to x (Fig. 16b). Both cannot be right. Which is?

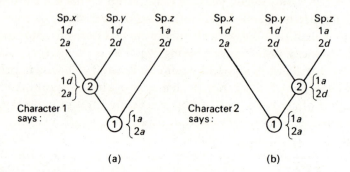

Fig. 7.16 Character state distributions producing a conflict of characters.

Three circumstances would lead to the above distribution: (1) errors in deciding which was the ancestral state of either character, (2) a hybrid origin of *y*, the species possessing the derived state for both characters, or (3) parallel or convergent evolution.

1. *If we were wrong* in deciding which were the ancestral and derived states of either character or both, then correcting this would remove the dilemma. Thus if we reversed the decision for only character 1, the state distributions would be:

	x	y	z
Character 1	a	a	d
Character 2	a	d	d

This distribution would give us the logical solution in Fig. 13b. If we were wrong as to the states of the other character 2 or of both characters, we would also have a resulting character state distribution removing the dilemma.

Realizing that correction of an error would solve the dilemma, at this point, we recognize that it behooves the investigator to reexamine the basis for deciding which is the ancestral state of each character.

2. *If species y was of hybrid origin,* it would be a mosaic type in which the character state of one species became incorporated with little or no change into either another species (introgression) or a hybrid species that arose as a third lineage from two parental ones. If in the hybrid, the character states were intermediate between those of the two parents, these states would not give the above tabulation of characters.

Whenever the two derived states of species *y* are additions rather than either one being a loss, a possible hybrid origin should be considered carefully. The techniques are outlined in a later section (p. 186).

3. *If parallel or convergent evolution* were involved, the question would be "which character represents a condition of the common ancestor necessary to explain the phylogeny, and which character evolved twice?" Three approaches are available in trying to decide this question.

a) Judging from the nature of the character states. In this there is little choice but to use a set of arbitrary truisms, drawn from well-studied examples of evolution in which a great deal of evidence has been available. The following appear to be the most useful, but the list is only a general one.

i) Losses occur many times more frequently than character novelties (Michener, 1949). Therefore if in Fig. 16 1*d* represents a loss and 2*d* a new character, the chances are much greater that Fig. 16b is correct.

ii) The more complex the character, the less likely that it evolved twice. If 1*d* is a simple addition whereas 2*d* is a more complex one, the chances are greater that Fig. 16b is correct.

b) Judging from many characters. If results from (a) are indecisive, it is necessary to search for more characters showing differences between the three species, to attempt to determine which conditions represent the ancestral and derived states, and to add these results to the tabulation. After this has been accomplished, it is sometimes possible to obtain a numerical assessment of the evidence which will give a tentative resolution of the phylogeny. This basis is a poor one because if many characters are needed, each will presumably give only inconclusive results.

c) Evaluating possible selection pressures. This approach involves an attempt to adduce selection pressures that might be acting on two related species so as to produce convergent or parallel evolution in one or more characters. Those characters considered the least likely to be so affected would be the more likely indicators of phylogeny.

For example, suppose that in a complex of three mammal species character 1 had the states thin hair (ancestral) and much thicker hair (derived), and that character 2 was represented by a curved penis bone or baculum (ancestral) and an angulate baculum (derived). Let us suppose further that species x and y lived in a colder climate than z. Mammals living in cold climates are under selection pressures normally leading to thicker hair, whereas there is no evidence that climate exerts any selection pressure on the shape of the baculum. Under these circumstances, the chances would seem better that

thick hair (1*d*) had evolved twice and that the angulate baculum
(2*d*) indicated the probable phylogenetic grouping.

Combinations of More than Three Species

When a group contains more than three species and each of the characters discovered exhibits two states, ancestral and derived, the computations outlined
for analyzing three species must be extended in the same fashion to cover the
additional species. When the group includes more species, the number of characters needed may increase, but as with three species, the minimum number of
characters needed will vary.

As in the three-species examples, the family tree is constructed by grouping together the smallest number of species sharing a distinctive derived character, then aggregating these smaller groups on the basis of more inclusive
shared derived characters and so on. The initial step is to tabulate the ancestral
and derived states for the species under study. A sample distribution for species *u* through *z* might be

	u	*v*	*w*	*x*	*y*	*z*
Character 1	*a*	*d*	*a*	*a*	*a*	*d*
Character 2	*a*	*a*	*d*	*a*	*d*	*a*
Character 3	*d*	*a*	*a*	*d*	*a*	*a*
Character 4	*d*	*a*	*d*	*d*	*d*	*a*

Species *v* and *z* share 1*d*, and can be grouped together; *w* and *y* share 2*d*, and
can be grouped together; and *u* and *x* share 3*d*, and can be grouped together.
This results in three pairs of species, each pair sharing a derived character. It is
then observed that the two pairs *w* + *y* and *u* + *x* share 4*d* and may be grouped
together, separate from *v* + *z*. When rearranged in this order, the phylogeny is
readily inferred (Fig. 17a).

With larger numbers of species, the same types of problems will arise as
with three species. For example, species *u* through *z* may have the following
distribution of character states:

	u	*v*	*w*	*x*	*y*	*z*
Character 1	*a*	*a*	*d*	*a*	*a*	*d*
Character 2	*d*	*a*	*a*	*a*	*d*	*a*
Character 3	*a*	*d*	*a*	*d*	*a*	*a*

Species *w* and *z* share 1*d*, *u* and *y* share 2*d*, and *v* and *x* share 3*d*, but no two
pairs share a derived character. The result based on this information will be
a logical trichotomy (Fig. 17b) in essence equivalent to that in Fig. 15b.

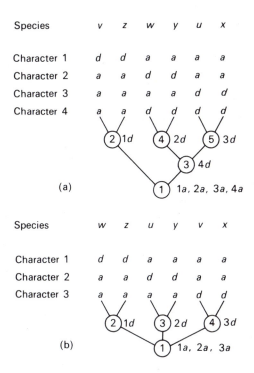

Species   v z w y u x

Character 1   d d a a a a
Character 2   a a d d a a
Character 3   a a a a d d
Character 4   a a d d d d

(a)

(2) 1d   (4) 2d   (5) 3d
(3) 4d
(1) 1a, 2a, 3a, 4a

Species   w z u y v x

Character 1   d d a a a a
Character 2   a a d d a a
Character 3   a a a a d d

(b)

(2) 1d   (3) 2d   (4) 3d
(1) 1a, 2a, 3a

Fig. 7.17 Phylogenetic solutions involving more than three species.

Phenoclines

Earlier, phenoclines were discussed as a possible means of determining the ancestral and derived states of characters. They may also be of unique importance in constructing family trees.

A phenocline is a set of sequential step-like states of a character that produce an orderly procession of changes from one extreme to the other. If the members of the phenocline form a distinctive group on the basis of sharing derived characters other than those involved in the phenocline, and if the ancestral state of the phenocline can be inferred, then the direction of the phenocline from ancestral to derived indicates the probable course of evolution. For example, let us assume a phenocline of four leaf shapes represented by four species set off as a unique group by derived character states of seeds, stems, and roots. If the leaf shape in species v is ancestral, those in species x,

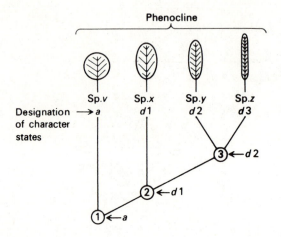

Fig. 7.18 A phenocline and its hypothetical ancestors.

y, and z would be a succession of derived states and the phylogeny shown in Fig. 18 would probably be correct.

In explaining the tree, it is convenient to designate the successive derived character states as $d1$, $d2$, and $d3$. The group ancestor (1) gave rise to species v, preserving the ancestral character state a, and to ancestor 2, in which derived state $d1$ evolved. Ancestor 2 gave rise to species x, preserving state $d1$, and to ancestor 3, in which derived state $d2$ evolved. Ancestor 3, in turn, gave rise to species y, preserving state $d2$, and to species z, in which state $d3$ evolved.

If each character state is represented by two or more species, we encounter the same logical problems as those explained in connection with Fig. 12c. An excellent example of this point is found in the caddisfly genus *Pycnopsyche*. The different character states of the lateral processes of the aedeagus (Fig. 19a) form a well-graded phenocline from a through $d5$. Comparing these states with conditions of the aedeagal arms in related genera, it is obvious that a is the ancestral state and that $d5$ is the most highly derived, with a progression from the long process and many short apical spines in a to a short process and only one elongate spine in $d5$.

Character states $d1$, $d3$, and $d5$ are each represented by 3, 2, and 8 species, respectively. On this basis it is tempting to construct the family tree in Fig. 19b, but this is logically incorrect. Ancestor 3 could have arisen from either ancestor 2 or from any of the lines leading to species 2, 3, or 4, just as was the case with species x in Fig. 12c. The same applies to ancestor 5 and its possible origin from either ancestor 4 or the lineages leading to species 6 or 7.

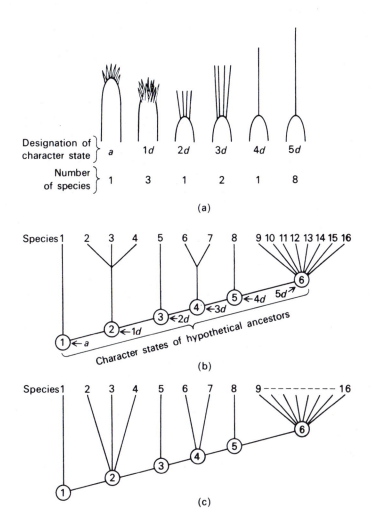

Fig. 7.19 A phenocline based on the aedeagal processes in 16 species of the caddisfly genus *Pycnopsyche* (a) and its phylogenetic interpretation (b,c). For explanation, see text.

Because ancestor 5 is represented by only one known species, ancestor 6 is portrayed logically.

To avoid misleading connotations, it is preferable to portray the circumstances as in Fig. 19c. If later it is shown that species 2, 3, and 4 share a derived state of a character that is ancestral in ancestor 3, and that species 6 and 7 share a derived state of a character that is ancestral in ancestor 5, this would then afford logical grounds for delineating the phylogeny as in Fig. 19b.

Deducing the Ancestry of Related Groups

Up to this point we have been discussing the procedures for determining intra-group phylogeny. The groups used were those defined by phenetic means at the beginning of the study (see p. 160). We now turn to the problem of inferring the inter-group phylogeny; that is, determining the blood relationships of each group to every other group in the system under study.

Inter-group study differs from intra-group study in one important point. In intra-group study, the inferences are made on the basis of character distributions among the *species* (see Fig. 17d). Theoretically, in inter-group study the inferences are made on the basis of character distributions among the *ancestors* of each group. This is outlined in Fig. 20a, a system of four phenetically clustered groups embracing a total of 17 species *a* to *r*. After their respective ancestors A1 to A4 have been defined by analyzing the character state distributions among the species comprising each group, then the next job is to figure out the phylogeny of these four ancestors.

In practice, the deduced ancestral states of ancestors 1 to 4 are usually expressions of different characters. The species group *a–d* may have been resolved on the basis of wing venation, hence ancestor 1 would be defined only in these terms; the species group *e–g* may have been resolved on the basis of male genitalia, hence ancestor 2 would be defined only in these terms; and in like fashion ancestors 3 and 4 might be defined respectively only in terms of color pattern, the distribution of body bristles, or other characters. At this stage it is necessary to define ancestors 1 to 4 in comparable terms that will afford a basis for analyzing them in the same way that we did species *a–d, e-g,* and so on. There are several ways to approach this problem.

First, if we compare the ancestral states of the characters deduced for ancestor 1, we may find that they are either ancestral or derived states with regard to those found in ancestors 2, 3, and 4. For example, in ancestor 1 the wing venation (character 1), although ancestral with regard to species *a–d*, might be highly derived with regard to the wing conditions adduced for ancestors 2, 3, and 4. In ancestor 3, the male genitalia (character 2) might be derived with regard to ancestors 2 and 4, but identical with those inferred for ancestor 1. This common derived character state distribution would give a definite phylogenetic grouping of ancestors 1 and 3. Similar types of character distributions in ancestors 2 and 4 might solve the phylogenetic position of these latter hypothetical forms. If, for example, the mouthparts (character 3) of ancestors 2 and 4 were ancestral with regard to their progeny but derived compared with ancestors 1 and 2, and the antennae (character 4) of ancestor 4 were ancestral with regard to their progeny, but derived compared with ances-

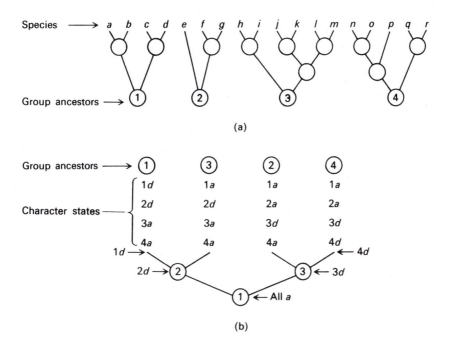

Fig. 7.20 Interpreting intergroup phylogenies.

tors 1–3, we would have the following tabulation of ancestral (*a*) and derived (*d*) *states*:

Group ancestors	1	2	3	4
Character 1	*d*	*a*	*a*	*a*
Character 2	*d*	*a*	*d*	*a*
Character 3	*a*	*d*	*a*	*d*
Character 4	*a*	*a*	*a*	*d*

These lead to the logical conclusion shown in Fig. 20b.

If it is impractical or impossible to pursue the problem in this fashion, the only remaining approach is to examine additional characters for which the ancestral and derived states in ancestors 1 to 4 can be inferred with a high degree of probability, and which show different states in the different hypothetical ancestors. Character states obviously ancestral for one of the groups or subgroups are those common to all its members. In such a situation it is necessary to discover character states common to one or two groups but not to others, and after this to establish which is the ancestral or derived state of any such character with regard to the overall group under study. This is often a difficult requirement that cannot be met with the evidence at hand. If so, a logical stalemate occurs that cannot be resolved until additional information comes to light.

Again resorting to experience, these stalemates are seldom encountered. Characters other than those previously considered often have states readily assignable to ancestral or derived, and with distributions that give highly probable clues as to the phylogeny of the ancestral forms. Frequently it is realized that ancestors 1 to 4 represent stages in a phenocline, which then leads to a rapid and highly probable resolution of the family tree.

If two groups prove to have identical hypothetical ancestors, then the combined progeny of both groups must be lumped together and reanalyzed according to the procedures of intra-group analysis. This requirement is a corollary of the situation depicted in parts (b) through (d) of Fig. 12. It may require a complete reexamination and reevaluation of the entire larger complex of species arising from the common ancestor.

Detecting Polyphyletic Groupings

The species analyzed phylogenetically by the foregoing procedures were segregated by various techniques that produced groups of relatively *similar* species but groups that were not necessarily *monophyletic*. The ever-present problem is to detect any groups that are polyphyletic. The only way to do this is to discover evidence indicating that one or more members of one group belong to another group. This is usually accomplished by noticing that a certain derived character state is shared by species previously placed in different groups. It is essentially the discovery of a conflict of characters between species of different groups. The logic in resolving the conflict is the same as that proposed for conflicts between species of the same group (p. 175).

Usually these inter-group conflicts arise when new characters become available, and the solution may be dramatic.

In some instances a derived character state might group one ancestor together with *one branch* arising from another ancestor. For example, in earlier groupings of certain caddisflies, several genera were considered a separate unit, the subfamily Polycentropodinae (Fig. 21a), of which the genus *Phylocentropus* was the earliest, primitive branch. Two other genera, *Protodipseudopsis* and *Dipseudopsis,* were considered a separate subfamily Dipseudopsinae based on enlarged pronotal warts in the adults. Later it was discovered that *Phylocentropus* and the Dipseudopsinae share many unique character states of the larval mandibles and legs. So complex were each of these that it was immediately highly probable that neither had evolved twice, and this recognition led to the

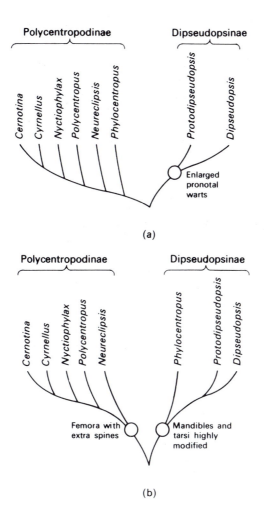

Fig. 7.21 Detecting incorrect initial groupings in caddisfly genera involving the sub-families Polycentropodinae and Dipseudopsinae of the family Polycentropodidae. For explanation, see text. (Redrafted from Ross and Gibbs, 1973.)

conclusion that *Phylocentropus* was the primitive member of the Dipseudopsinae (Fig. 21b).

Thus when new derived character states are discovered in one group, it is necessary to examine these same traits in related groups and in groups perhaps considered only distantly related. Only in this fashion can we unearth errors of previous phenetic groupings and of previous phylogenetic conclusions. Thus the process of detecting more and more inclusive sets of taxa based on shared derived characters becomes more effective with a wider knowledge of the taxa related either closely or distantly to the group under study.

Detecting Hybrid Origins

As indicated in the section concerning character conflicts, the character state distribution

Species	x	y	z
Character 1	d	d	a
Character 2	a	d	d

immediately invites consideration that y may be a hybrid species combining the derived character states of x and z. The reasoning behind this inference is that in plants a number of hybrid species of this type have been produced artificially by laboratory hybridization followed by careful selection. Such an explanation removes the necessity of postulating that one or the other derived character states evolved independently twice. This latter postulate is not difficult to defend if the derived states are simple losses, but is open to serious question if the derived states are additions.

If the species are not amenable to genetic experimentation, one circumstance indicates strongly the likelihood of the hybrid origin of a species. If the derived characters combined in the suspected hybrid species are the ends of different phenoclines, each occurring in a different group, there is an extremely high probability that the species is of hybrid origin. The simplest character state distribution indicating this circumstance is

Species	x	y	z	h	a	b	c
Character 1	a	$d1$	$d2$	$d2$	a	a	a
Character 2	a	a	a	$d2$	$d2$	$d1$	a

The logical phylogenetic tree based on this distribution is shown in Fig. 22. In nature instances approaching this have been found in the leafhopper genus *Erythroneura* (Fig. 23).

In this setting, it is highly improbable that two characters would progress through identical steps twice and produce almost identical character states in nonhybrid populations.

Numerical Phylogeny

Numerical methods for constructing phylogenies have been proposed by Camin and Sokal (1965) and Throckmorton (1968b), who have developed computer programs for making the necessary calculations. Both proposals stress the need to score all utilized characters as to ancestral and derived states, and to score character states so that they reflect presumed evolutionary direction.

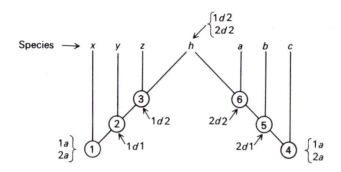

Fig. 7.22 Detecting species of hybrid origin. For explanation, see text.

The logic used in these proposals follows in general that explained above in this chapter. As Camin and Sokal pointed out, these methods have as yet no technique for dealing with hybridization. They also do not attempt character weighting based on complexity, or on gains versus losses, both of which have a bearing on the probable answers to situations involving conflict of characters. They have, however, developed some tests for phylogenetic parsimony and for testing this against judgments concerning decisions as to the ancestral and derived states of individual characters. It is certain that improvements in numerical phylogeny will be made which could facilitate the investigation of groups having a large number of taxa.

Fitch and Margoliash (1967) used a numerical method based on cytochrome c amino acid sequences in 20 selected organisms ranging from molds to man. They calculated the minimum numer of nucleotides that would need to be altered in order that the gene for one cytochrome would code for another; this number they termed the *mutation distance* between any two cytochromes. By comparing these numbers, they arranged the 20 species in a numerical relationship. At the level of phyla, their results are in accord with conclusions based on morphology; in the various groups of vertebrates, there are numerous instances of discordance with relationships based on morphology. This is not surprising when one considers that they are basing their conclusions on a single character, even though it exhibits many states. The same situation often arises when a phylogeny is attempted on only a single morphological character. The discrepancies suggest that biochemical characters also exhibit convergent or parallel evolution, unequal rates of evolution in different lineages, problems in establishing ancestral and derived character states, and other difficulties encountered in the study of any type of character.

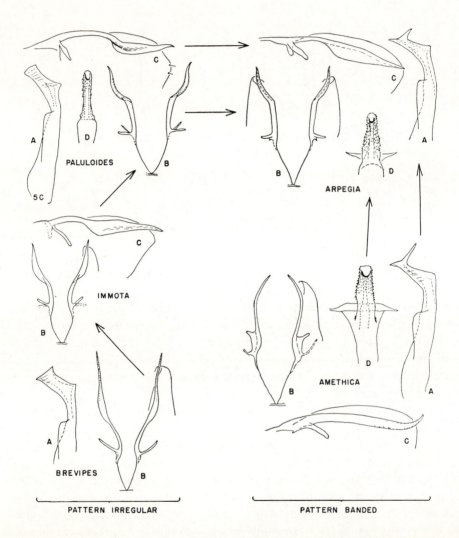

Fig. 7.23 Diagnostic features of five species of *Erythroneura*. *E. brevipes, E. immota, and E. paluloides* belong to the brevipes branch of the *lenta* complex; *E. amethica* belongs to the *trivittata* complex; *E. arpegia* is the presumed hybrid combining the derived pygofer hooks (b,c) of *E. paluloides* with the derived banded pattern and aedeagus (d) of *E. amethica*. The style (a) is ancestral in *E. amethica* and *E. arpegia*, derived in the others.

Polyploid Bisexual Species

Polyploid species are superimposed on the isoploid family tree according to their similarity to various species represented in that tree, combined with the various chromosome numbers possessed by each species. Three general situations are encountered.

1. *Series of polyploids having multiples of the same gametic number, e.g., 12, 18, 24.* This situation is exemplified in Fig. 24. Let us assume that we have a species $m=12$ that has sorted to the same cluster of species as a, b, and c with gametic numbers of 6, 6, and 7, respectively, on the basis of other characters. Species m could be either an autopolyploid of a or an allopolyploid of a x b. If m is remarkably like a in character states and exhibits no traits characteristic of b, it is almost certainly an autopolyploid of a. If an additional species $n=18$ also resembled a but not b, it would be the product of a x m, usually considered an autopolyploid (Fig. 24a).

If m has some traits of a and some of b, it is probably an allopolyploid of a x b (Fig. 24b).

2. *Series of polyploids having irregular higher gametic numbers, e.g., 11, 15, 26.* If in the initial clustering process such a series sorted out character-wise with a group of species $a=5$, $b=6$, $c=7$, and $d=8$, they would probably be allopolyploids because none is a doubling of one of the isoploid chromosome numbers. Species $m=11$ would presumably be the allopolyploid of c x d; species $n=15$ would presumably be the allopolyploid of c x d; species $o=26$ would be the allopolyploid of m x n (Fig. 24c).

If two of the basic isoploid species have the same chromosome number, as in Fig. 24d, problems arise as to choice of parentage. In the case of $m=11$, it could have arisen from either a x b or a x c. If m combined traits of a and b, it presumably arose from a x b; if it combined traits of a and c, it presumably arose from a x c. Should the chromosomes of the isoploid species involved have distinctive shapes or other features, this cytogenetic evidence can often provide virtual proof of the actual parentage. On occasion suspected parents can be crossed in various combinations and polyploid mutations of the hybrids induced with colchicine; if the result resembles the polyploid species in question, its ancestry is virtually established.

A third type of possibility is also found on occasion, exemplified by a situation in which m had certain traits of a, none of b, but other traits not found in the known species of the group. The two suggested solutions would be that m was either an allopolyploid of a and a species in another group, or an allopolyploid between a and some as yet undiscovered species of the same group. For example (Fig. 24e), if species $q=19$ in certain traits resembled n,

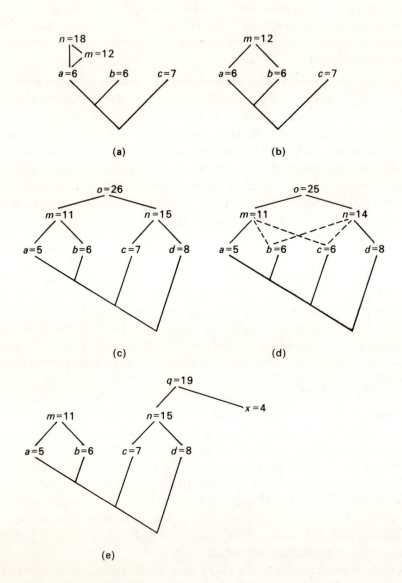

Fig. 7.24 Deciphering species of polyploid origin. For explanation, see text. Numbers are gametic chromosome numbers.

it would presumably be an allopolyploid between n and some as yet unknown species $x=4$.

3. *Isolated polyploids.* Compared with its relatives, a species may have a sufficiently high chromosome number to be considered an almost certain polyploid, and it also may not sort out with any of the species clusters in the group. This situation calls for a comparison of the traits of the suspected allopolyploid with species in all the species clusters in the group and also with species in other groups. Intergeneric allopolyploids are well known in plants, indicating the breadth of search that might be required to obtain clues as to the ancestry of the species thus isolated character-wise.

Inserting polyploid species into the family tree is thus seen to be an educated juggling act requiring a thorough knowledge of character states both within and outside the immediate study group and the exercise of considerable ingenuity.

Apomict Species

The only logical way to interject apomict species into the family tree is first to prepare such a tree on the basis of bisexual species in the group, then add the apomicts to that tree on the basis of morphological or other character similarities. When such a tree is available, this procedure will result in adding isolated lineages here and there in the family tree. In many groups such as the bacteria, viruses, and the Fungi Imperfecti, there is as yet no phylogenetic framework based on bisexual species into which the apomicts can be fitted. In these instances the only recourse is to attempt to relate the apomict entities with each other and with bisexual species by comparing the states of whatever characters are available.

EXPRESSING PHYLOGENETIC CONCLUSIONS

A family tree is a hypothetical construct based on the application of the scientific method to a collection of sensed data, and resulting in the investigator's concept of the path of evolution in the group under study. To be amenable to testing by other scientists, the family tree must be explained clearly and fully. The following items are a minimal requirement for clarity:

1. List the taxa included in the study and the characters on which the initial group was differentiated from others.

2. List the intra-group characters used and their various states, plus directions as to methods of preparation needed for adequate observation.

3. Explain the postulated ancestral and derived states of each character, and the reasons for these conclusions.

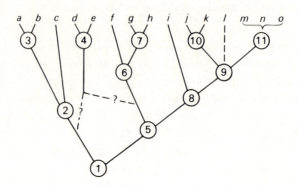

Fig. 7.25 Several conventions for indicating doubtfully related taxa in a family tree.

4. Explain the postulated events for each branch or fusion in the family tree. This is most easily done by designating each hypothetical ancestor with a distinctive number or letter, defining each ancestor character-wise, and stating the implied happenings between each ancestor and its immediate known progeny.

5. If different characters led to conflicting conclusions, explain why the proffered solution was chosen.

6. If some lineages can be placed in the tree on only a tentative basis, indicate the doubt by either a dangling line, a broken line, or a question mark (Fig. 25).

With these points expressed clearly and fully, the family tree becomes a scientific instrument inviting further testing, investigation, and extension.

8

Phylogeny, Fossils, and Time

An enticing possibility inherent in a phylogeny or family tree is to express character change and phylogenetic branching in relation to geologic time. When the family tree is worked out, both branching and character changes are apparent. The next question is: When did these different events happen? If we knew this for all the lineages, we would have a time scale for our tree and could correlate changes in specific characters with time. We could then ascertain rates of evolution for individual characters or some measure of the total character changes occurring at different periods in the evolution of the lineages. A complete correlation of character change, phylogeny, and time would give us a vista of the entire panorama of life evolving through the ages, from the 3–billion year old dawn unicells to present day daisies and man.

The backbone of correlating the evolution of life with geologic time is the fossil record. An identified, dated fossil can do something that nothing else can do: Tell us the minimum age of a group or lineage. For example, the archaic fern-boring sawfly genus *Blasticotoma* was known for years only as a living species in Europe. Its geologic age was unknown. Then a fossil wing from the Oligocene Florissant shales of Colorado was identified as *Blasticotoma*. This find immediately established the age of *Blasticotoma* as *at least* 40 million years; it takes 40 million years out of guesswork and puts it into the realm of objective observation.

The fossil record provides information on the past evolution of life and the only clues as to when the major groups of organisms existed. The earliest known fossils are forms resembling rod-shaped bacteria in South African deposits dated as 3.1 billion years old (Kenyon and Steinman, 1969). The next younger extensive set of fossils comprise a variety of unicells in the Gunflint shales of Ontario, Canada, dated as 1.6 billion years old (Barghoorn and Tyler, 1965). The first extensive fossil biotas begin in the Cambrian period of the Paleozoic era about 570 million years ago. From that time to the present there is a substantial record for every geological period (Fig. 1). It is far from

193

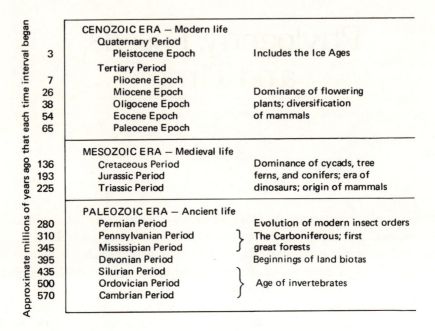

Fig. 8.1 Geologic timetable from the early Cambrian to the present and the major biological events during this period. (Dates adapted from B. Kummel, 1970).

complete, nowhere complete for the whole geologic sequence, and admittedly scanty for certain groups or certain time intervals.

The record is best for marine organisms. Especially in the Paleozoic era, epeiric seas covered vast areas of the continents. Remains of the marine life settled to the bottom and became fossilized, over the eons resulting in fossiliferous limestone or shale deposits often thousands of feet thick. At the present time many of these old sea beds have been elevated, and the fossils can be found in quarries and roadcuts, or found on eroded hillsides. For the Paleozoic this record is remarkable for many groups of marine organisms. It is much less extensive for many periods of the Mesozoic and Cenozoic eras, but nevertheless surprisingly good sequences of general types have been preserved and discovered.

For terrestrial forms the record is much more spotty, especially for small, fragile organisms or parts. The best fossil record is of the hard parts of animals and of the stems, leaves, pollen, and spores of plants. An unusually good record of both kingdoms, including insects, is contained in the Coal Measures or Carboniferous of the Paleozoic. Insect fossils are especially abundant in Permian strata, the uppermost beds of the Paleozoic. For much of the

Mesozoic and early Cenozoic, deposits containing fossils of the more fragile species are infrequent. Notable examples of these include: amber inclusions, especially rich in insects and small plant parts, that have been found in a few Cretaceous and mid-Cenozoic deposits; and fossils found in shales resulting from deposits in small lakes, such as the Oligocene shales at Florissant, Colorado, and other localities in western North America. Fossils of almost recent forms are found in abundance in bogs and other deposits formed during the Pleistocene.

For the Mesozoic and early Cenozoic, when presumably the primitive angiosperms and many of the more modern terrestrial groups of invertebrates were evolving, few fossils of flowers or soft-bodied animals have been found. This circumstance has hindered the evolutionary and temporal analysis of these groups. In spite of these and other problems inherent in the fossil record, it has contributed a vast amount of information concerning the timing of evolutionary events (Kummel, 1970).

DATING FOSSILS

The more precisely and reliably fossils can be dated, the better is the evidence they offer for correlating phylogeny with time. Most fossils must be dated by determining the age of the stratum in which they are found. In a relatively few instances, a fossil may be dated by measurement of its own characteristics.

Dating Fossils by Their Strata

Most fossils are found in the same beds in which the fossil was originally formed. The tremendous beds of fossiliferous limestone contain the fossil remains of dead organisms that sank to the bottom of the sea and eventually formed an integral part of the limestone. Fossils found in the Florrisant shales of Colorado are the remains of plant and animal parts that were washed into a turbid Oligocene lake, became embedded in layers of fine mud settling on the floor of the lake, and were later converted into impressions in the resultant shales. In these instances, if one determines the age of the stratum in which the fossil occurs, one also determines the age of the included fossil.

Some fossils become redeposited. Bits of amber, concretions, coal balls containing fossils, may be eroded out of the strata in which they were originally formed and deposited in another layer of sediments. Hard, solid fossils such as trilobites, pieces of petrified wood, or bones may be redeposited in the same fashion. Under these circumstances, the fossils are found in strata younger than those in which they formerly occurred. Excellent examples are the Baltic amber fossils of Europe and the Canadian amber fossils found in existing lakes and along present-day lake shores.

The redeposited nature of the fossils is usually easy to spot by a number of well-known geological techniques. Once this point has been established, any and all geological clues must be sifted through in an effort to determine the strata in which the fossils were originally formed. When this has been done, the problem of dating the fossils is that of dating the parental strata.

Of the methods available for dating strata, the following three are most used by paleontologists.

1. *Stratigraphic dating.* Large areas of the continents comprise a series of deposits or bands of sediments superimposed one on top of another. These reflect part of the history of the earth's crust and are the basis for the geologic timetable. The fossils embedded in these strata give us our picture of the evolution of life and its diversity through time. There are many irregularities and breaks in the record of these strata progressing from place to place, leading to much difficulty in correlating strata of one locality with those of another.

Before the advent of more precise dating methods, the age of strata was based on estimates of the length of time required for the deposits to have accumulated. Because of the admitted guesswork inherent in these calculations, the only certain age determination was that under normal processes of deposition, a stratum was older than the one above it and younger than the one below. Time estimates on this basis are termed relative dating.

2. *Radiometric dating.* When rocks or sediments are formed, sometimes certain radioactive elements become locked in them. As the elements decay to some stable chemical, the decay products also become incorporated in the sample along with the remainder of their parent element. Thus the radioactive elements, thorium and uranium, decay through a complex series of changes to stable lead. The rates of decay of the various radioactive forms are known, and further there is excellent evidence that this decay rate has been constant since the elements were first formed. On these premises, if one measures the amount of parent material and the amount of decay products in a given sample, one can calculate the age of the formation of the sample. An account of the methodology in using these methods is given by Rutten (1962) for the longer-lived elements and by Broecker (1965) for the shorter-lived ones.

A dozen or more radioactive elements are employed in radiometric dating, ranging from radioactive carbon (C^{14}) with aging possibilities up to 70,000 years to some of the thorium and uranium isotopes with aging possibilities up to many billions of years. For marine strata in the Cenozoic, potassium–argon (K–A) ratios have provided excellent radiometric dating.

Both relative and radiometric dating methods present many problems, chief of which is the unfortunate circumstance that, except for C^{14}, samples suitable for radiometric dating are often not directly associated with fossil-bearing strata. This is especially true for the Paleozoic. An accounting of these difficulties and their attendant pitfalls is given by C. A. Ross (1970).

3. *Biotic assemblages.* Concerning both age-dating and interlocality correlation, one of the greatest difficulties is the fact that terrestrial deposits are few and are often only partially preserved. The forces of erosion and glaciation have obliterated vast areas of these deposits, amplifying to a staggering degree the problem of correlating the small bits and pieces remaining on a scale satisfactory for stratigraphic correlation. In the cases where radiometric dating is inconclusive, paleontologists have worked out a scheme of species assemblages that appears to express the evolutionary state of entire faunas at different periods of time. Newly discovered finds are compared with various standard assemblages already described and defined (Fig. 2). If certain fossiliferous beds in one locality occur in a superimposed relationship, the relative time relationship of these assemblages is obvious. Correlating isolated faunal assemblages from different regions or continents is less certain. Difficulties associated with younger (Cenozoic) assemblages were detailed by Repenning (1970); those for older (Paleozoic) assemblages by C. A. Ross (1970).

There are several other methods available for dating strata. Two of the most promising are determinations of past positions of the poles, and past reversals of the earth's magnetic field, based on evidence from magnetic and crystallization patterns in certain types of iron deposits. As more information becomes available from these two fields of inquiry, and especially as it becomes integrated with other geologic phenomena, we can expect more accurate dating of fossil-bearing sediments.

Dating Individual Fossils

Two methods are employed. If the fossil itself contains measurable amounts of radioactive isotopes, these can be measured and the fossil dated accurately. The chief organisms so far aged by this method are woody fossils of plants preserved relatively intact in Pleistocene beds, and calcium carbonate shells. In these instances, dating the fossil has *ipso facto* dated the stratum also.

The second and highly intriguing method involves the correlation of daily growth increments in organisms such as marine coral fossils and from this obtaining a value for the number of days in a year. Astronomical evidence has shown that the earth's rate of rotation has been decreasing, resulting in a decrease in the number of days per year (Wells, 1963). The days per year have been calculated for various geologic periods (Fig. 3). If the days per year can be determined from the fossil, a comparison with this chart will show its age.

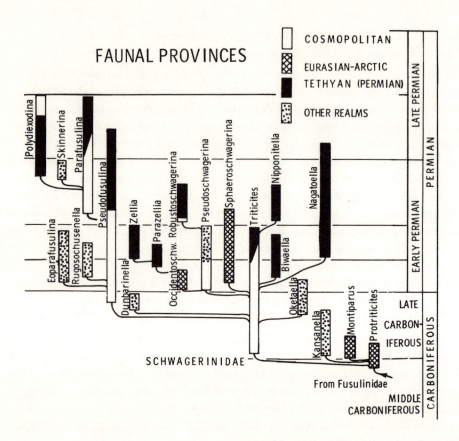

Fig. 8.2 World distribution of some genera of the fusulinacean family Schwagerinidae during late Paleozoic time. Horizontal groupings of genera give faunal assemblages characteristic of various times. (From C. A. Ross, "Concepts in late Paleozoic correlations," *Geologic Society of America*, Special Paper, 124:7–36. 1970. Reprinted with permission of the Geological Society of America and the author.)

In practice, several or all of these methods may be combined in determining as accurately as possible the age of a particular stratum and the fossils it contains.

CONVERTING PHYLOGENIES TO TIME

In many groups only recent species are known, and for these only a limited number of time phenomena can be determined. The same is true of groups all of whose taxa are known from a single time level. Both situations present the limitations of determining time relations when only contemporaneous spe-

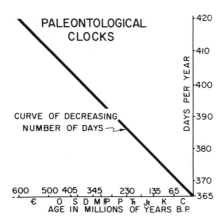

Fig. 8.3 Reduction in number of days per year as calculated from astronomical data and corroborated by paleontological evidence. (From C. A. Ross, "Concepts in late Paleozoic correlations," *Geologic Society of America*, Special Paper, 124:7–36. 1970. Reprinted with permission of the Geological Society of America and the author.)

cies are involved. For groups having a sequence of fossil forms representing various time levels, more incisive time relations can be established.

Interpretations from Contemporaneous Taxa

Two criteria can be used for correlating phylogeny with time — sequential branching of the family tree and the equation of character difference with geologic time.

Sequential Branching

By use of the family tree alone, it is possible to indicate only the order in which certain events occurred prior to the time level of the species studied. Thus in the family tree depicted in Fig. 4, ancestor 1 occurred before ancestor 2, and 2 occurred before 3. Similarly, ancestor 1 occurred before ancestor 5, 5 before 8, and 8 before 9 or 10. It is irrelevant which time level is involved. The taxa under study could be all Devonian, all Triassic, all Recent, or all in any other time level.

With a family tree like the one in Fig. 4, one almost automatically considers that the ancestors at each level were synchronous, leading to the assumption that ancestors 3, 4, 7, 9, and 10 all occurred at the same time and that ancestors 2, 6, and 8 occurred together at an earlier time. Such an assumption would appear to be plausible especially if the species of each pair differed by about the same amount of character change — in other words, if taxon *a*

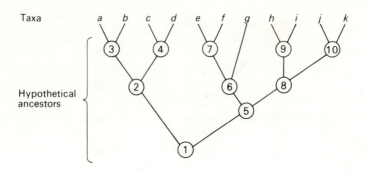

Fig. 8.4 Diagrammatic family tree illustrating sequential branching.

differed from *b* about the same amount of character change as *c* differed from *d, e* from *f,* and so on. Because we have excellent evidence that different lineages may evolve at different rates, however, we logically can make horizontal time correlations of inferred hypothetical ancestors only on a tentative basis. For example, even if they differed by the same amount, *a* and *b* might have evolved their differences at a faster rate than did *c* and *d,* in which case ancestor 3 would be younger than ancestor 4.

If we apply Simpson's (1944) analysis of rates of change in phylogenetic lines to the situation in Fig. 4, we can arrive at tentative time correlations for the relative ages of the hypothetical ancestors. Simpson discovered that, in a particular group and in a given period of geologic time, the majority of the lines evolve at a medium rate, and a relatively small number evolve at much higher or lower rates. On this basis, the chances would be good that most of ancestors 3, 4, 7, 9, and 10 were contemporaneous. By the same reasoning, ancestors 2, 6, and 8 would have a good chance of being contemporaneous.

Character Differences

As explained above, although various measures of character difference may be calculated between different successive points of the tree, the rate of change cannot be converted with a high probability into units of geologic time. Only various measures of the relative rate of change can be made. For example, if in Fig. 4 taxon *a* were theoretically almost identical with ancestor 3 whereas taxon *b* exhibited great differences from it, then the logical conclusion would be that, since the time of ancestor 3, taxon *b* had evolved at a more rapid rate than *a*.

Air et al. (1971) proposed a method for dating the time at which groups diverged from their common ancestors, using differences in amino acid sequences in globin proteins, including myoglobin and alpha and beta hemoglobins. For their calculations, they assumed that amino acid differences, counting each insertion or deletion as one difference, occurred at the same rate in each lineage. Thus in Fig. 4, if there were 14 differences between taxa *a* and *b*, *a* and *b* would have been considered as differing equally from ancestor 3 by a count of 7 amino acid differences. Using data from 12 assorted species of vertebrates ranging from the carp to humans, they arrived at a result comparable in general to the recorded or surmised history of the vertebrates.

In spite of this, certain criticisms of the method arise. In the first place, using similar evidence from studies of cytochrome *c*, Fitch and Margoliash (1967) found evidence of different rates of protein evolution in different lines. Their result is in keeping with morphological data that gives evidence of little change in certain lineages yet fantastic change in others. Examples include: the club mosses, some lines of which are remarkably similar from the Carboniferous to the present, yet whose immediate ancestors gave rise to others that evolved into the most complex flowering plants; and the wingless bristle-tail insects whose living species must be little different from pre-Carboniferous ancestors, yet some of whose ancestors eventually evolved into the most complex insects. Actually the rates of evolution given by Air et al. (*op. cit.*) differ markedly in the three proteins, although the standard deviations are high. The authors discussed several other interesting paradoxes. In the second place, the date of origin taken for several eutherian groups is highly simplified and may have introduced time errors not of the authors' making.

In spite of these criticisms, the method certainly opens up definite possibilities as a means of estimating the times of phylogenetic divergence. As its authors pointed out, few protein series of this type are yet available; many more are needed to obtain a test of the application of the method.

Geographic and Ecological Correlations

If certain geographic or ecological correlations can be established between a group of contemporaneous species and other groups having an earlier fossil record, some of the evolutionary events concerned with the group of interest can be correlated with geologic time. First, however, the phylogeny of the group must be explored with regard to geographic and ecological phenomena. These aspects are discussed in Chapters 9 and 10.

INTERPRETATIONS FROM FOSSIL SEQUENCES

If fossils of a group are known from various time levels, many more inferences can be made concerning geologic time and the family tree. The comprehensiveness and reliability of these inferences are proportional to the information content of the fossil record. To be of maximum use in phylogenetic interpretation and temporal placement, a fossil must preserve enough characters to allow its placement in the family tree of its group and must be accurately dated in regard to geologic time. The former point was the major concern in the previous chapter; the latter point is the major concern in this chapter.

If the fossil record were complete, we would have few problems in correlating time and phylogeny. Such a record would require well-dated fossils, in excellent condition, of all the species existing in every subdivision of geologic time. The first benefit of such a record would be its aid in deciphering problems of phylogeny. Attendant benefits would include the possibility of knowing when phylogenetic branching occurred, how long each lineage persisted through time, and the type and rate of morphological change within the lineage through time. The fossil record, however, is far from complete, resulting in many difficulties concerning both phylogeny and geologic time and the correlation of the two.

In spite of the difficulties encountered, practically all well-preserved fossils have been placed somewhere in the phylogenetic scheme. Because of the limited number of characters they exhibit, many fossils can be assigned only tentatively with regard to their relatives, but a surprising number have been placed with a high degree of probability. Most of these fossils can be related to some living forms. The giant protozoans comprising the order Fusilina are known only from Paleozoic and Mesozoic strata, but details of their anatomy indicate that they are a branch of the protozoan class Foraminifera, of which many species are living today. A recently discovered Pennsylvanian cycad proved to belong to the plant order Cycadales and to be remarkably similar in cone structure to the present-day genus *Stangeria* (Taylor, 1969).

The large number of these past and present associations is quite remarkable when you consider that the associations must be based on characteristics that can be seen on dead specimens. No other attributes will contribute to a comparison of fossils with living species. The extent of this knowledge is an especial tribute to the ingenuity of paleontologists in devising various techniques for exposing fossil structures.

To avoid letting time relations obscure phylogenetic inferences, it is desirable first to work out the family tree embracing all the taxa in the group from all time horizons, including the present. The result will be a tree in the general form of Fig. 4. After that we can indicate above each taxon its position in geologic time, as has been done in Fig. 5. For simplicity, this example supposes that each taxon is known from only one geologic time period. The next step is to transpose this tree to a time chart in which each species is placed in its proper time level, but in which the phylogenetic arrangement is unaltered (Fig. 6a).

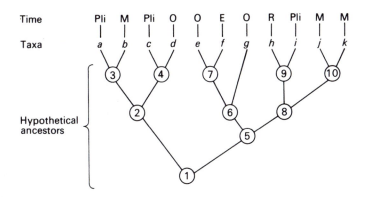

Fig. 8.5 Diagrammatic phylogeny shown in Fig. 8.4 on which time data have been super-imposed.

Figure 6a is strictly provisional pending a comparison of each taxon with its immediate ancestor. The ancestors are initially placed lower than any of their offspring taxa; the distance below is no measure of geologic time, but is solely an expression of relative time. When the older taxon of a fork is like the ancestor in *all* observed character states, it replaces the hypothetical ancestor and may be considered as the parent of the younger one. Thus in Fig. 6b, if *b* is like ancestor 3 in all observed character states, then it is assumed that *b* is ancestor 3 and that *a* arose directly from *b*. In this example, *b* and *a* would be sequential species.

If the younger taxon is like the ancestor, it indicates that the older taxon arose from the younger one. This situation is further described in Fig. 6c. If *a* is like ancestor 3, then it is assumed that *a* is ancestor 3 and that *b* arose from *a*. This situation requires that *a* existed prior to the known record of *b*.

Both of these examples may be combined in the same complex. For example, if in the *e*, *f*, and *g* complex, *f* = ancestor 7 and *g* = ancestor 6, then each of these two taxa would replace its respective ancestor and give the solution shown in Fig. 6d.

If *a* and *b* each differ from Ancestor 3 in at least one different derived character state, then their relation will be that shown in Fig. 6a.

If in the solution of Fig. 6b the three character states of ancestor 2, *b*, and *a* were steps in a phenocline, they would be a *chronocline*. A chronocline is a phenocline in which each successive character state occurs in one of a series of sequential species. Chronoclines are extremely useful in following the evolution of a lineage, as exemplified by Matthews' (1970) analysis of Pliocene beetles (Fig. 7).

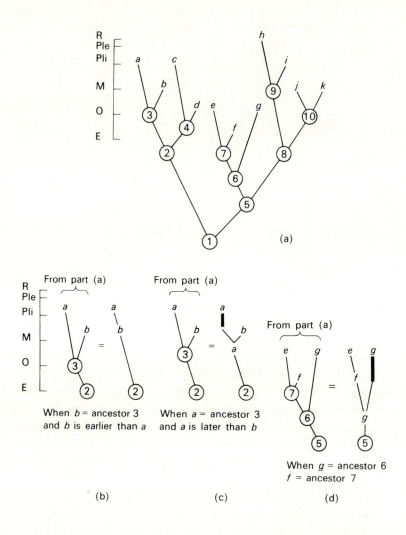

Fig. 8.6 Adjustment of phylogeny shown in Fig. 8.5 with geologic time. For explanation, see text.

From the circumstances portrayed in Fig. 6 it is impossible to make any time determinations for hypothetical ancestors not identified as sequential species. This is not always the case. If a hypothetical ancestor occurs subsequent to a geologically dated sequential species, it may be associated with some measure of time. For example, if the phylogeny and time dispersal of species *a*, *b*, and *c* were as shown in Fig. 8a, and species *c* were identified as ancestor 2,

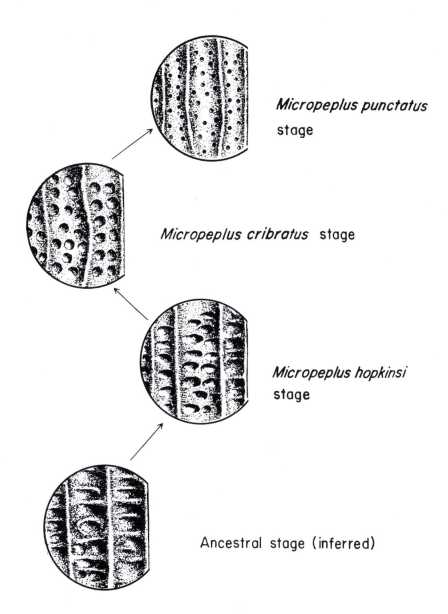

Micropeplus punctatus stage

Micropeplus cribratus stage

Micropeplus hopkinsi stage

Ancestral stage (inferred)

Fig. 8.7 Chronocline inferred for the evolution of staphylinid beetle elytra leading from an inferred ancestral stage, through the Pliocene species *Micropeplus hopkinsi* and *M. cribratus* to the Recent *M. punctatus*. (From J. V. Matthews, Jr., "Two new species of *Micropeplus* from the Pliocene of western Alaska with remarks on the evolution of Micropeplinae (Coleoptera: Staphylinidae," *Canad. J. Zool.*, **48**, 779–788, 1970. Reprinted with permission of the Canadian Journal of Zoology.)

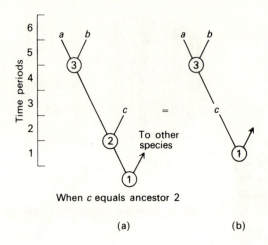

Fig. 8.8 Time relations of certain hypothetical ancestors.   For explanation, see text.

then the adjusted phylogeny would be that in Fig. 8b. Ancestor 3 would then be dated as occurring at some time between time periods 3 and 6.

These basic situations can be applied to the taxa and hypothetical ancestors of any group.

Geologic Ranges, Phylogeny, and Time

The *geologic range* of a taxon (whatever its rank) is the span in geologic time between its earliest fossil record and its latest known occurrence. The latter may be the present or Recent. If all we know concerning a group of taxa is their respective geologic ranges, we have no means of inferring any more about the time of occurrence of any of the taxa involved than is shown by the graphic representation given in Fig. 9a. This record could be essentially correct or woefully inadequate, but from these data alone we could make no other meaningful estimate of probable time values.

If the phylogeny of the taxa has been worked out, at least one additional possibility of time judgment may be available. Should the phylogeny be of the pattern shown in Fig. 9b, in which daughter lines arose sequentially from preceding ancestral forms, then the geologic ranges of the later taxa would appear to approximate the real time ranges of these forms. Should the phylogeny be as in Fig. 9c, then it is apparent that each lineage could be much older than its geologic range would indicate. How much older is in no way apparent. This or some part of its measure will have to await new fossil finds bearing directly on the topic.

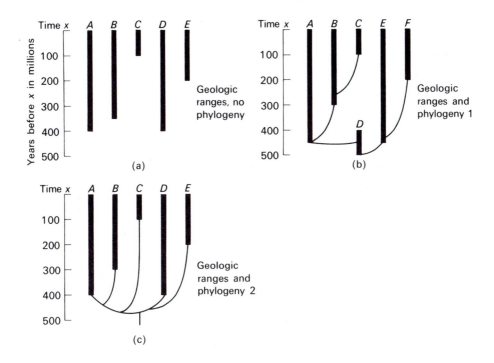

Fig. 8.9 Approximations of minimum and maximum ages of taxa *A* through *E* represented in the fossil record. For explanation, see text.

Fossil Representation

The more fossil species we have scattered over the entire phylogeny, the better we will be able to associate all its branching with geologic time and to learn the rate of change through time of the characters possessed by the species included.

It is remarkable, however, how much information can often be gained by the discovery of only one or two fossil species that can be placed definitely in the family tree. Frequently the reliable dating of one branch will indicate the possible correlation of earlier or later parts of the tree with certain geologic events, or give other types of time clues for exploration. An example is afforded by the caddisfly family Glossosomatidae. The branch comprising the tribe Agapetini contains about 75 Recent species and one extremely well preserved Baltic amber fossil, *Electragapetus scitulus*. The latter can be fitted exactly into a detailed family tree of the tribe (Fig. 10), indicating that the lines evolving into *Catagapetus* and *Agapetus* originated in pre-Oligocene time.

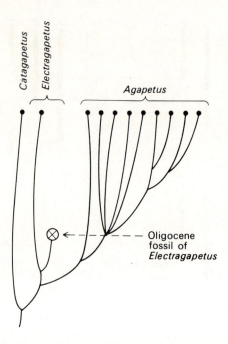

Fig. 8.10 Simplified phylogeny of genera and species groups in the caddisfly tribe Agapetini, showing the position of its only known fossil.

9

Geographic Dispersal

After the species of a group have been determined, their phylogeny adduced, and this phylogeny correlated with geologic time, the next step in systematic exploration is trying to discover and explain the distribution over the globe of the various lineages of the group.

EVIDENCE OF GEOGRAPHIC CHANGE

The geographic distribution of living things has intrigued systematists ever since the 15th- and 16th-century voyages of exploration brought to light what to European eyes were strange and wonderful plants and animals. As classifications and collections improved, both botanists and zoologists tried to bring some order to the burgeoning wealth of distributional evidence. The first person to realize that comparative distribution patterns of organisms gave clues concerning worldwide dispersal patterns seems to have been Sclater (1858), who based his conclusions on the distribution of birds. Sclater, however, was an antievolutionist who envisaged these patterns as static rather than dynamic. Wallace was struggling with this same problem when he independently hit upon the idea of natural selection as the causal agent directing the course of evolution. After he and Darwin presented their joint paper on this topic, Wallace continued his studies on the distribution of animals, and in 1876 arrived at his set of biogeographical realms (Fig. 1).

Wallace's "realms" expressed chiefly the similarities of distribution of animal taxa in various areas of the world. These biogeographic realms give a measure of the areas of the world that have each had a long and independent evolution of their biota, resulting in overall taxonomic differences between the biotas of the different areas. The realms do not in themselves contribute directly to the dynamic idea of dispersal, the movement of species from one part of the world to another.

The idea of dispersal arose when attention was turned to drastic geographic discontinuities existing between the ranges of closely related taxa.

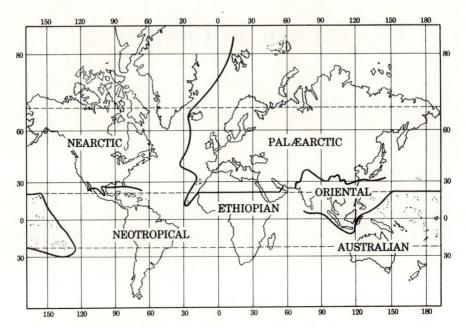

Fig. 9.1 Faunal realms or regions of the world. (Modified from Sclater and Wallace, in H. H. Ross, *A Textbook of Entomology*, 2d. ed., John Wiley & Sons, New York, 1950. Copyright © 1950, by John Wiley and Sons, Inc., and reprinted with permission.)

For example, the salamander family Cryptobranchidae has only two living genera, *Megalobatrachus* in eastern Asia and *Cryptobranchus* in eastern North America. It is obvious that sometime in the past the common ancestor of these two genera dispersed between eastern Asia and eastern North America. The mammal family Procyonidae has two genera, *Procyon* (raccoons) and *Nasua* (coatis), in the Americas and the genus *Ailurus* (cat-bear) in the Himalaya. The common ancestor of these animals must also have spread between North America and Asia. The primitive frog suborder Amphicoela contains only the genera *Liopelma* in New Zealand and *Ascaphus* in western North America. These two genera indicate quite a remarkable dispersal of their common ancestor from almost one pole to the other. The plants have their share of similar disjunct distribution patterns. For example, the primitive tribe Cladothamnae of the heath family Ericaceae comprises only three genera, *Elliottia* in southeastern United States, *Cladothalmus* along the Pacific Coast of North America, and *Tripetaleia* in Japan. At least one ancestral form of this tribe must have dispersed between North America and Asia.

Examples abound of similar disjunctions between distinctive species in the same genus. The plant genus *Diphylleia* has two species, one in the southern Appalachians, the other in Japan. The plant genera *Rhododendron, Lirio-dendron, Sassafras,* and many others have species in eastern Asia, others in eastern North America, and none existing in the areas between. Homologous *Hibiscus* genomes have been found in South America and Africa (Menzel and Martin, 1971). Insects exhibit similar disjunctions between almost any two continents — Africa vs. South America, New Zealand vs. South America, and so on.

Many discontinuities occur within the range of the same species, as discussed in Chapter 5. The skunk cabbage and other examples cited there give abundant evidence that the ranges of many species are fragmented into two or more widely separated units.

These examples are intercontinental. Marine taxa exhibit many examples of disjunctions that indicate former connections between distinctive ocean areas now separated by land barriers.

Ekman (1935, 1953) has cited examples of marine animals; Bartholomew et al. (1911), de Beaufort (1951), and Darlington (1957), of many other animals; and Cain (1944) and Dansereau (1957), of plants. Croizat (1958) has reviewed many of these plus added his own examples for both plants and animals. These authors list many examples involving Recent and fossil relatives such as the Tsetse fly *Glossina,* whose only known living species occur in Africa, but of which Oligocene fossils are known from the western United States.

If the propagules of these groups were extremely vagile and could disperse haphazardly to all quarters of the earth, the range disjunctions would need no further explanation. But in the examples cited above, plus myriads of others on record, this is not the case. There is no known natural way by which a terrestrial frog could disperse now between New Zealand and the North American Northwest or by which a *Liriodendron,* plant or seed, could disperse between eastern Asia and eastern North America without a relatively continuous stretch of habitats suitable to each species between the now-disjunct areas.

The information from these disjunctions therefore proves beyond doubt that in the past there has been a tremendous dispersal of phylogenetic lines between various parts of the world, many of them requiring land or water connections that are not now in existence.

The wealth of examples indicates also that dispersals having these requirements occurred between practically every combination of continents and at many different periods of geologic time. Under these circumstances it is obvious that life has had a complex pattern of dispersals and discontinuities requiring continental and oceanic conditions different from those existing at present.

There are two complementary methods of investigating the geographic history of life. One involves using the data from species of organisms to obtain a basic knowledge of the probable dispersals that have occurred in various lineages of plants and animals. The second entails the probable avenues of dispersal existing at various geologic times. Obtaining an insight into this parameter involves the highly controversial matter of a reconstruction of the changing conditions of the continents and oceans and their climates as they existed in the past. To date both methods are in a state of exploration, with many discrepancies appearing between them. When the two are more fully understood, one should complement the other, and the combination should provide a comprehensive view of the geographic history of life.

NATURAL DISPERSAL PATTERNS

The special data used in these dispersal studies are the localities from which individual collections have been made. Ideally, the records for each species, Recent or fossil, are plotted on maps as accurately as possible. If a species is known from more than one period of geologic time, records for each time level are kept separate. If species phylogenies are not available, then the distribution of genera or families is plotted.

Dispersal by Man

If the dispersals involve widely separated populations or subspecies of the same species, or if some of the distinctive species in the group are cultivated by man, there is the possibility that they were transported accidentally or knowingly by man. If so, the area of natural occurrence of each species is in doubt and may pose serious problems in establishing the pattern of natural dispersal. For example, various game species of animals have been introduced into almost every continent on the globe. The same is true of large numbers of cultivated plants. Since the time of the first transoceanic travel, species of plants and small animals have been unwittingly transported back and forth between continents and a large number of species have succeeded in becoming established in an alien country. These include many weeds, parasites or pests of crops and livestock, and an assortment of other non-noxious species that were in soil, straw, or other material being taken across the ocean. Recently many species involved in biological control have been introduced into continents far from their native home.

For many species of importance to man, the time and direction of dispersal has been noted in the literature. These instances pose no problem to the student of natural dispersal. But for the "unimportant" species that do not attract attention and for prehistoric movements of cultivated species and others associated with them, there is no published record. Specialized types of sleuthing are needed to track down these dispersals. An extensive literature outlines methods and research problems encountered in tracing the origin and dispersal of cultivated plants (De Candolle, 1884; Vavilov, 1951; Anderson, 1952; Merrill, 1954). Interesting examples of sleuthing methods used in tracking down the original homes of noneconomic species are given by Lindroth (1957). If circumstantial evidence does not give a probable answer, phylogenetic-geographic studies of the dispersal of the entire group to which the individual species in question belong will frequently indicate instances of transport by man.

At this point it is important to determine, if possible, which widespread ranges are due to man's activity and which are natural.

Dispersal patterns have three primary facets: direction of specific dispersals, number of dispersals and disjunctions in a given taxon and time of specific dispersals. Each facet presents its own peculiar problems and will be discussed separately.

Direction of Dispersal

Four principal methods have been employed in solving this question: numerical, ecological, phylogenetic, and fossil.

Numerical

In both plant and animal biogeography, the idea is widespread that the area containing the largest number of species is the area of origin. Cain (1944) pointed out that this is by no means universally true of plants and Ross (1962) the same for insects. Because this idea is not a truism, even though true in many instances, it cannot be accepted as a valid assumption for scientific analysis.

Ecological

It has been argued that if the direction of dispersal of one group of lineages was considered probable, then any other taxon associated with it ecologically would have had the same direction of dispersal. When considered further, this maxim could be logically fallacious; hence it is something that must be demonstrated rather than be assumed. Two hypothetical examples illustrate the problem. First, let us assume that two ecologically similar but taxonomically different kinds of plants A and B lived in Asia and North America, respectively, and that an avenue occurred allowing both to disperse between the two continents. After the dispersal, it would be impossible to determine from the later widespread geographic ranges of the two species what had really been the history of their dispersals. If it were determined that A had originated in Asia, but no such history were determined for B, it would not be apparent where B originated.

Second, let us assume a producer-consumer relationship such that *A* consumes only *B*. If it were determined that *B* lived originally in Asia, then spread to North America, it would be logical to assume that *A* also originated in Asia and spread to North America along with its host *B*.

These hypothetical examples illustrate the point that geographic data alone may or may not be able to resolve the probabilities of using the ecological method in deducing direction of dispersal. Kind and degree of ecological relationships must also be considered, and these are discussed in Chapter 10.

Phylogenetic

This third method has the possibility of answering at least some of our questions. It is, however, a demanding mistress with a rigid list of logical "do's and don'ts."

The general idea of this method was first outlined by Kinsey (1930, 1936) as a result of his historic studies on the gall wasp genus *Cynips*. He proposed that if the phylogeny of a group were known, and if the geographic distribution of the species were superimposed on this phylogeny, probable inferences could be drawn concerning the geographic dispersal of various phylogenetic lines. He made such interpretations for the various living species groups of *Cynips* (Fig. 2). Kinsey has been criticized for attempting such conclusions without fossil evidence, but his method is completely sound in that it must be used even if fossils are involved.

The phylogenetic method of inferring direction of dispersal consists in associating geographic differences of the taxa with successive forks in the family tree. Information from only one fork can give a clue only concerning areas between which dispersal occurred. In the frog suborder Amphicoela, with the genus *Liopelma* in New Zealand and *Ascaphus* in western North America, it is obvious that both arose from a common ancestor that eventually gave rise to these two genera. But this single bit of present-day distribution must be the result of the division of the range of this common ancestor into at least two units, the eventual evolution of one into *Liopelma* and the other into *Ascaphus*. Where did the common ancestor arise? It could have been in New Zealand, North America, or almost any other place on the globe. All we know for certain is that there must have been some as yet indecipherable dispersal of either the common ancestor or some of the evolutionary stages of *Liopelma* and *Ascaphus* such that one is now in New Zealand, the other in North America. To obtain results concerning the origin of the ancestral form and direction of dispersal, we need more information.

The power of this additional information is shown in the case history of the *Wormaldia kisoensis* complex of caddisflies. At first only two species were recognized in the complex, *W. kisoensis* from Japan and *mohri* from eastern North America (species 1 and 2 of Fig. 3a). These two species plus

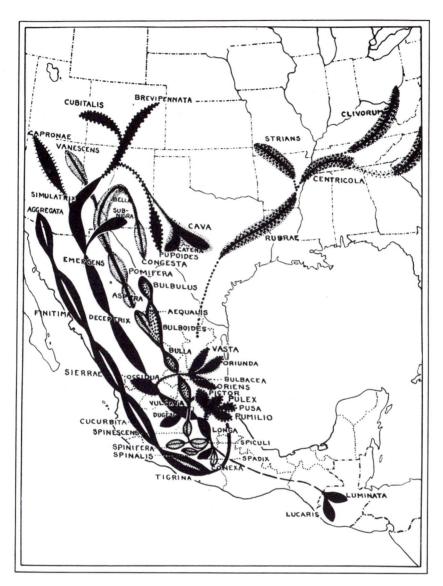

Fig. 9.2 Phylogeny and distribution of the *atrusca* group of the gall-wasp genus *Cynips*. (From Kinsey, 1930)

their distribution indicated an intercontinental dispersal of their common ancestor but gave no hint as to the direction of dispersal. It could have been from east to west, or west to east, with a 50-50 chance for each.

Later a species 3 from the Philippine Islands came to light, with phylogenetic relationships shown in Fig. 3b. On the basis of Occam's razor (the principle of least assumptions), this arrangement gives a 2-1 chance that ancestors 1

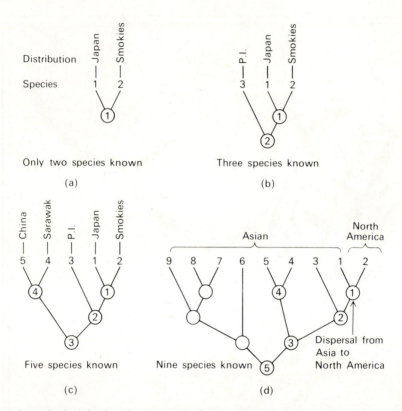

Fig. 9.3 Steps in deciphering the dispersal of the *mohri–kisoensis* group of the caddisfly genus *Wormaldia*. For explanation, see text.

and 2 were both Asiatic, and that the direction of dispersal was from Asia to North America. This requires only one intercontinental dispersal to explain the phylogeny and distribution. Any other arrangement requires two intercontinental dispersals. For example, an explanation considering ancestor 2 to be American would require the assumption of two America-to-Asia dispersals, one by ancestor 2 giving rise to species 3 and another by ancestor 1 giving rise to species 2.

Still later, species 4 from Sarawak and species 5 from China were added to the complex, with the phylogeny and associated distribution shown in Fig. 3c. Here ancestor 4 seems to be indubitably Asiatic, indicating a greater probability that ancestors 1, 2, and 3 are also. Later, when species 6, 7, 8, and 9 were added, with phylogeny and distribution as shown in Fig. 3d, it was practically certain that the entire complex evolved in Asia except for one branch of ancestor 1 that dispersed to North America, as outlined in Fig. 4 (Ross, 1956).

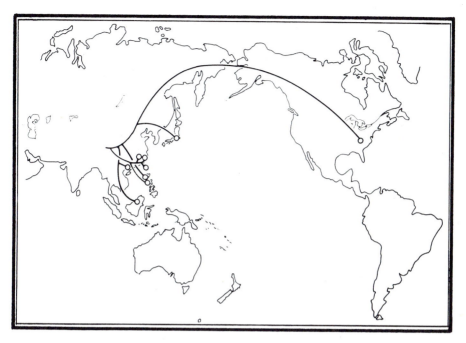

Fig. 9.4 Map showing phylogeny and distribution of the species treated in Fig. 3. The circle in Japan represents *Wormaldia kisoensis*; that in eastern North America, *W. mohri.* (From H. H. Ross, *A Synthesis of Evolutionary Theory,* © 1962. Reprinted by permission of Prentice-Hall, Inc., Englewood Cliffs, N. J.)

In reviewing this example, it is evident that the larger the number of successive ancestors indicating the same origin of a lineage, the greater is the probability that the assumed point of origin is the correct answer, which forms a firmer basis for inferring the direction of dispersal of related lineages. It must always be remembered that additional species added to the system, either Recent or fossil, may give new evidence in conflict with previous conclusions.

Mechanics of a dispersal

An intercontinental dispersal such as that of ancestor 1 in Fig. 4, resulting in two distinct species, involves a series of events:

1. Ancestor 1 evolved as a distinct daughter species of ancestor 2, through one of the various types of speciation outlined in Chapter 5. Its distribution was probably in northeastern Asia (Fig. 5, first stage).

2. Because of climatic changes to the north, or genetic changes in the species (discussed later in this chapter), ancestor 1 spread into northern North America, presumably as far east as the Appalachian region (Fig. 5, second stage).

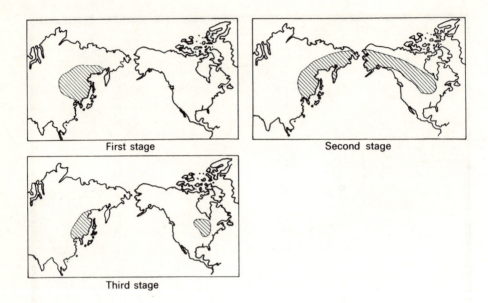

First stage   Second stage   Third stage

Fig. 9.5 Steps in the intercontinental dispersal and disruption of a species: First stage, original range on one continent; second stage, dispersal to another continent; third stage, extinction of middle populations.

When in later figures, an ancestor is termed "intercontinental," the reference is to this event of the series.

3.   Changes in ecological conditions, probably a colder or drier climate and/or the erosion of hilly terrain due to glaciation, caused an extinction of widespread ancestor 1 in the more northern parts of its range. This would have resulted in two or more isolated populations of that species, at least one in Asia and one in eastern North America (Fig. 5, third stage).

4.   Once isolated, each population evolved into distinctive species. The only two of which we have a record are *W. kisoensis* in Japan and *W. mohri* in the Great Smoky Mountains of southeastern North America.

The dispersal of ancestor 1 and its subsequent evolution into geographically distant daughter species therefore required an initial dispersal of a range, then its disjunction into isolated segments, then the evolution of these segments into distinct species. *Wormaldia kisoensis* and *mohri* are together an excellent example of the end products of this process.

An excellent example of stage 2, a widespread intercontinental species, is afforded by another caddisfly, *Glossoma intermedium*. It now occurs from

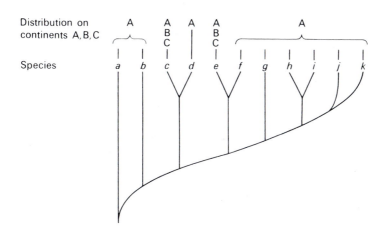

Fig. 9.6 A common type of distribution pattern. For explanation, see text.

Scotland in western Europe, across Eurasia to Alaska, and eastward to Michigan and Missouri of the United States. The Missouri populations constitute a glacial relict remnant isolated from more northern populations; other populations may also be isolated. A warming of post-Pleistocene climates should result in an effect comparable to that of the *Wormaldia kisoensis-mohri* example.

A frequent outcome of the phylogenetic-geographic analysis is illustrated in Fig. 6. All the species of the group are restricted to one continent except one or two that occur also on others. It is obvious that the evolution of the entire complex occurred on continent A, and that subsequently species *c* and *e* became widespread. This situation immediately raises the question: Did these intercontinental dispersals occur naturally or through human transport? The answer will require some recognition of the possibilities of human transport. For example, Fig. 6 fits the case of the mosquitoes *Culex pipiens* and *C. fatigans* (corresponding to species *c* and *e*) which are worldwide in temperate and tropical climates; all their close relatives (corresponding to species *a-b* and *f-k* occur in Africa. It is known that (1) populations of each species from Africa and other areas are identical morphologically, (2) both species breed with great success in water barrels, and (3) the sailing vessels of many centuries carried their water in open barrels. In view of these circumstances it seems almost certain that these two wideranging species of *Culex* were spread by sailing vessels from an original African home to all parts of the world. Similar instances involving no discernible connection with human transport would imply a natural dispersal.

Number of species

There is a current belief among some biogeographers that the direction of dispersal will be more apparent in groups represented by a larger number of species. That this is not so is well illustrated by the mosquito genus *Mansonia*. In this genus, the greater the number of species considered, the less is the likelihood of inferring the home of the common ancestor (Fig. 7). In this example, if only the species arising from ancestor 3 are considered, it is obvious that this ancestor arose and all but one of its daughter species evolved in the Old World. The North American species arising from ancestor 4 represents a dispersal of part of ancestor 4 from the Old to the New World. Ancestors 5 and 7 and their progeny evolved in the New World and ancestor 8 and its progeny evolved in the Old World. But when we add together all the species within the genus, we lose the possibility of inferring the origins of these combined systems. The two ancestral forms involved, ancestors 2 and 6, obviously had an intercontinental dispersal but where each arose before it became widespread we do not know. Either one or both could have originated in either the New or Old World and later dispersed to the other. As a result, we have no basis for inferring the area of origin of ancestor 1, the progenitor of the genus, from information obtained only within the genus. As a matter of interest, this is all that is available at present bearing on this question.

Fossil

If specimens are available from different levels of geologic time, there is a widespread belief that the locality of the oldest fossil is the area of origin. This may indeed be the case, but unless it is supported by phylogenetic evidence there is no good reason for assuming this to be so. Referring again to the sawfly genus *Blasticotoma* (Fig. 8), it is impossible to determine objectively where the genus evolved on the basis of its one known fossil from North America and its present distribution restricted to Europe. If the Oligocene species gave rise to the European species (Fig. 8a), the former could very well have been a widespread intercontinental species arising originally in either the Old or New World. If both species shared different derived characters, their phylogenetic relations would be as depicted in Fig. 8b, which indicates that each arose from an earlier ancestor having ancestral states of the characters involved. Thus at one time intercontinental species could have originated either in the Old or the New World.

Abundance of fossil records may not provide a more probable answer. A case in point is the genus *Ginkgo,* at present known only from eastern Asia. From middle Jurassic to the Miocene, the genus was Holarctic. Then (Fig. 9) its range was constricted to its present Asiatic distribution (Tralau, 1968). Its earliest known fossil occurrence is a single lower Jurassic record from Russia. Considering its later extensive range, Tralau concludes that this one early Rus-

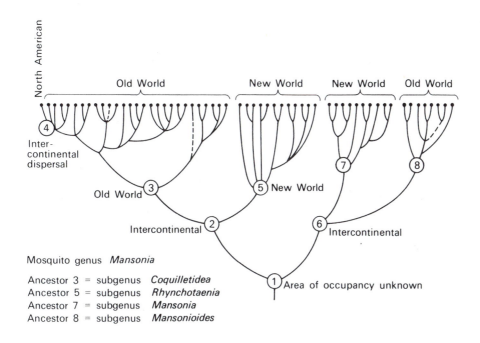

Fig. 9.7 Simplified family tree of the mosquito genus *Mansonia*.

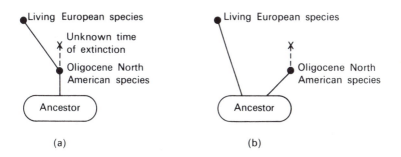

Fig. 9.8 Possible relationships of the single Recent European species and the North American Oligocene species of the sawfly genus *Blasticotoma*.

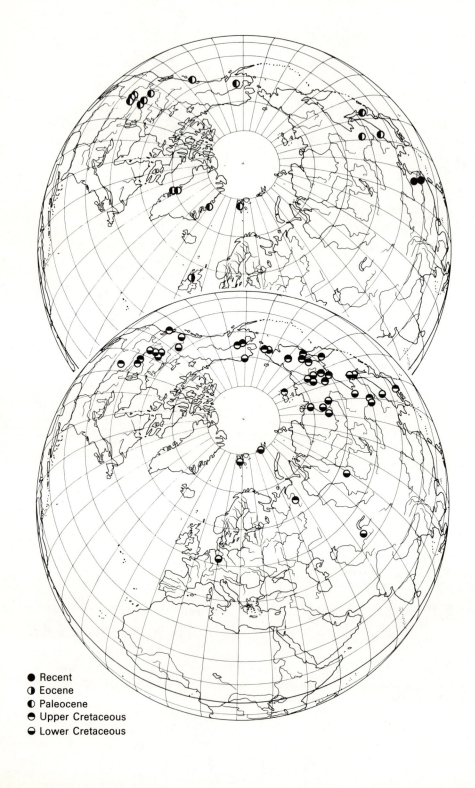

● Recent
◐ Eocene
◑ Paleocene
◓ Upper Cretaceous
◒ Lower Cretaceous

sian fossil is, at best, weak evidence for an Asiatic origin of the genus, and that the total fossil evidence gives no information concerning migration routes of *Ginkgo* during its early migration history.

If the fossil record included a series of serially related and more ancestral species on one continent with derived species on another, comparable to the situation in Fig. 10, the probable direction of dispersal would be clearly indicated.

Number of Dispersals

If a group has one or more species in one region but only one species in another, this constitutes evidence for inferring only one interregional dispersal. Such is the situation in the *Wormaldia kisoensis* complex (Fig. 3), with several species in Asia and only one in North America. On the evidence of geographic evidence alone, a single dispersal will best explain the data.

If a group has more than one species in each of two regions, there is the possibility of more than one dispersal. The only known method to determine the probable number of dispersals is first to decipher the phylogeny of the species comprising the group, then plot geographic distribution on the family tree. Inferences can often be made on the basis of this arrangement of information. In the group of four species in Fig. 11a with *a* and *b* in North America, *c* and *d* in Asia, the most probable solution is that ancestor 2 was North American, ancestor 3 was Asiatic, and that both arose from one intercontinental dispersal of ancestor 1. If the geographic ranges were different, more dispersals might be indicated. For example, if the ranges were as in Fig. 11b then two intercontinental dispersals would be necessary to account for the circumstances. In neither case would the original home of ancestor 1 be apparent. In systems with more species, more dispersals might be indicated. In the Holarctic leaf-

Fig. 9.9 Known range of the tree genus *Ginkgo* at selected horizons from Lower Cretaceous to the present. Note that the genus, once widespread, is now known to occur naturally from only two localities in eastern Asia. (From H. Traulau, 1968.)

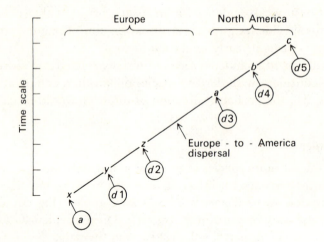

Fig. 9.10 Diagrammatic situation involving six species *x*, *y*, *z*, and *a*, *b*, *c*; their character states are indicated in the circles. For explanation, see text.

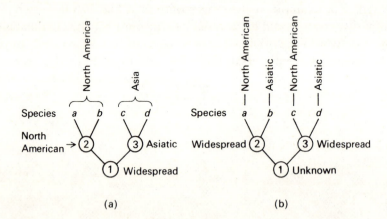

Fig. 9.11 Illustration of relationship between phylogeny and distribution with number of dispersals.

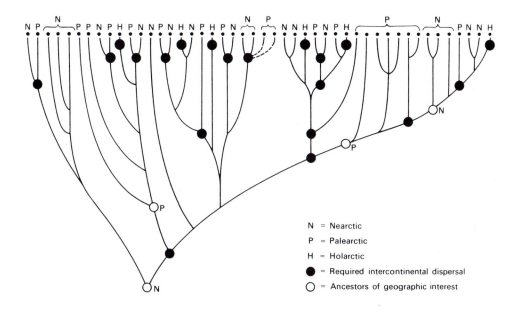

Fig. 9.12 Phylogeny of the leafhopper genus *Macrosteles* indicating probable intercontinental dispersals (larger black dots) and ancestors of geographic interest (open circles). The smaller upper black dots represent the known species. (Modified from T. E. Moore and H. H. Ross, 1957.)

hopper genus *Macrosteles* (Fig. 12), a minimum number of 19 intercontinental dispersals is required to explain the distribution of its 45 species (Moore and Ross, 1957).

The caddisfly genus *Phylocentropus* illustrates the necessity of combining data from known species from all time levels. There are seven Recent species whose phylogenetic relationships are well documented, and an eighth species *shigae* from Japan that cannot as yet be placed satisfactorily in the family tree. If only the seven better known Recent species are considered (Fig. 13a), a single dispersal from North America to Asia would satisfy the known phylogeny and geography. It so happens that four species of *Phylocentropus* have been described from Baltic amber fossils of presumed Oligocene age, and that these species can be correlated phylogenetically with the Recent species. When a revised phylogeny is prepared and correlated with geography (Fig. 13b), it is immediately certain that at least three intercontinental dispersals of the genus occurred, one each in the *auriceps, placidus,* and *lucidus* groups, with Eurasia as the more probable original home.

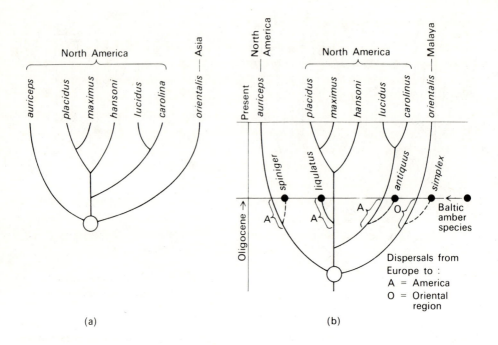

(a)   (b)

Fig. 9.13 Phylogeny, distribution, and inferred probable dispersals of the caddisfly genus *Phylocentropus*: (a) on the basis of only the Recent species; (b) on the basis of both fossil and Recent species. The Japanese species *P. shigae* is unplaced. For explanation, see text.

In case the examples from *Macrosteles* and *Phylocentropus* introduce the idea that the greater the number of species in a group, the greater the number of interregional or intercontinental dispersals, we should refer back to Fig. 7. In this system of 58 species, its geographic data can be satisfied by inferring only three intercontinental dispersals.

Time of Dispersal

The question explored here is: When did specific dispersals occur? The inquiry embodies many of the considerations surrounding the question of Chapter 8: When did specific phylogenetic branchings occur? The two questions are closely related because wide dispersals usually result in phylogenetic branchings, although the converse is not necessarily true.

Concerning time of dispersal, combinations of circumstances fall into the three following categories.

Family tree inferred, no fossils known

Under these circumstances, ordinarily only the sequence of dispersals within the phylogenetic scheme can be determined. Thus in Fig. 7, the dispersal of ancestor 3 occurred before that of ancestor 4, the two being sequential branches on the same lineage. But this tree alone does not support the contention that ancestors 2 and 6 dispersed at the same time. Perhaps they did, but not necessarily so. No correlation with radiometric time is possible with only these data.

Under certain circumstances such a time correlation can be made. If the dispersal pattern of one taxon having no correlation with geologic time can be correlated with that of another taxon that is correlated with geologic time, the time of dispersal of the latter taxon indicates the probable time of dispersal of the former. Thus in the leafhopper genus *Erythroneura* it is evident that the large Nearctic fauna arose from two or three dispersals from Eurasia to North America. From evidence only from the leafhoppers which have no known fossils, the time or times of these dispersals cannot be inferred.

Their shrub, vine, or tree angiosperm hosts, however, have a marvelous fossil record. It indicates that during mid-Cenozoic times these hosts formed a Holarctic belt that became fractured after mid-Cenozoic time into their present isolated European, Asiatic, North American, and other fragments. From these data of distribution plus the group phylogeny, it seems evident that the intercontinental dispersal of *Erythroneura* occurred during mid-Cenozoic when their hosts had a trans-Holarctic distribution (Chaney, 1940).

No detailed family tree inferred, some fossils known

The situation usually fitting this category is when one or a few Recent species and a few fossils of the same genus are known, and when the fossil specimens do not exhibit characters needed to relate them in detail to the living forms. An example is the sawfly genus *Blasticotoma* (Fig. 8), known from one Recent European species and an Oligocene species from western North America, the latter identified only by a wing showing the venation highly characteristic of the genus (Benson, 1942). It is evident that an intercontinental dispersal is involved, but its time of occurrence is doubtful. If both Recent and fossil specimens are the same species, the dispersal could have been before, during, or after the Oligocene. If the Recent species is a sequential one arising from the fossil one, the dispersal would have been after the Oligocene (Fig. 8a). If both are distinctive species that arose from a previous ancestor, then the dispersal was before the Oligocene (Fig. 8b). But with no knowledge of the phylogeny, a choice between these alternatives is impossible.

Family tree inferred, including fossils

In these circumstances, the same type of maximum and minimum age determinations can be made for branchings associated with dispersals as explained
in Chapter 8. In the simplified example in Fig. 14, with supposed fossils *x*
and *y* placeable in the phylogeny as indicated, the Asia-to-America dispersal
of ancestor 3 occurred at some time *after* the existence of species *x*; hence the
Miocene is the maximum age of this dispersal. The intercontinental dispersal
of ancestor 1 occurred *before* the existence of species *y*; hence the Oligocene
is the minimum age of the dispersal.

In the genus *Phylocentropus* (Fig. 13b), only the minimum ages can be
given for all three dispersals, which is pre-Oligocene.

ROUTES OF DISPERSAL

When the direction, number, and time of dispersals have been adduced with a
high degree of probability, there remains the question: By what route did each
dispersing lineage reach its new home? As explained earlier in this chapter, the
evidence indicating past dispersals of lineages now represented by widely separated disjunct populations or species also implies that in the past the geography and/or the climates of the world have been different from their present
state.

As a result, it follows that we cannot interpret past dispersals inferred
from neontological or present day evidence on the basis of routes suggested
by the present geography and climates of the globe. The difficulties thus
raised vary with individual dispersls and are proportional to our knowledge of
the circumstances surrounding each example.

These difficulties involve not only inferences based on neontological species, but also all those based only on contemporaneous species, whatever their
age. If sufficient well-preserved fossils were available from successive strata,
both the dispersal route and time would be determinable after the group phylogeny had been adduced. But if this idealistic situation did not prevail, a
knowledge of probable dated avenues of dispersal would be helpful in answering two questions: What route did the dispersing lineages follow? and When
did they do it?

Questions relating to avenues of dispersal for continental biotas are different in several aspects from those concerning marine biotas. It is therefore
convenient to deal with the two separately.

Continental Biotas

These include both terrestrial and aquatic forms. The latter are tied to continents just as surely as the former. Water lilies, pond weeds, frogs, freshwater

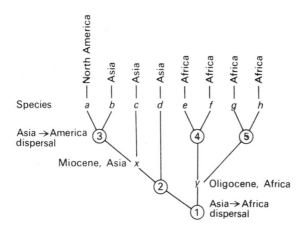

Fig. 9.14 Illustration of fossils and their relationship in dating dispersals. For explanation, see text.

fish, and all but a handful of aquatic insects cannot exist in salt water for any extended period, usually a matter of a few days at most, and hence cannot complete their life cycles except in the continental setting.

In many intracontinental instances the route of dispersal seems evident. For example, phylogenetic evidence indicates that certain lineages of winter stoneflies dispersed in the past between the Appalachian Mountain area to the Ozark-Ouachita Mountain area in Arkansas and Oklahoma. Various types of evidence suggest a Pleistocene time. Geologic evidence demonstrates that during this period there existed a corridor of artesian country extending from the Appalachians, across southern Indiana and southern Illinois, to the Ozarks of southern Missouri which merge with those of Arkansas. During cooler glacial periods this stretch would have provided a series of spring-fed streams perfect for the intermountain dispersals of *Allocapnia* (Ross and Ricker, 1971). Many other intracontinental dispersals can be explained as readily.

Intercontinental dispersals present more difficulties. Taking a simple case, Chaney (1940) found that in mid-Tertiary many trees of the temperate deciduous forest extended across all Eurasia and North America. These data can best be explained on the basis of one or several intercontinental dispersals of each taxon between Eurasia and North America. On the basis of present continental positions, it would appear obvious that the route of these dispersals was via the Bering Straits bridge or Beringia, between northeastern Asia and northwestern North America. But according to one school of thought, western Europe was connected to eastern North America at this mid-Cenozoic time and the Bering Straits bridge had not yet formed. If this latter view were correct, the dispersal route would have been via the western Europe–eastern North America route.

From these examples it is clear that our knowledge of avenues of dispersal is dependent to a large degree on geological evidence indicating the topographic and climatic conditions during various past periods of earth's history.

Intercontinental Avenues

Since early in the nineteenth century, geologists have recognized that the earth's crust was dynamic or changing. The presence of strata obviously laid down originally in shallow areas but now occurring in raised mountain chains convinced early nineteenth century geologists that part of the crust had become elevated in some fashion. When the continents became well mapped, geologists were intrigued by the remarkable fit of the west coast of Africa and the east coast of South America. This coincidence and other data led Wegener in 1912 to postulate that these two continents were formerly one, that the supercontinent later broke up, and that the present continents drifted apart. Wegener (1924) envisaged that about 300 million years ago all the continents formed a single mass that since has broken up into the present continents (Fig. 15).

Wegener's theory of continental drift was rejected by many scientists. To explain patterns of terrestrial animal distributions, Matthew (1915), relying chiefly on evidence from the vertebrates, proposed a model of successive north-to-south dispersals following present-day land connections. The model did indeed explain a large number of data, but required the assumption of a highly complex set of northern extinctions. Others who disagreed with Wegener's general hypothesis nevertheless recognized that paleontological evidence indicated different continental associations in past times. This circumstance was indicated strongly by the distinctive *Glossopteris* flora of the Permian, which appeared to have evolved at about the same time in Africa, Australia, Tasmania, southern India, and South America. To account for this distribution, geologists hypothesized that these areas once formed a single continent, Gondwana, parts of which were thought to have later become ocean. Schuchert's 1924 proposal of the geography of early Permian time (Fig. 16) is an excellent example of these ideas.

Later, Hilgenberg (1933) and Carey (1958, 1970) proposed that the continents were indeed becoming more and more separated, but due to an expansion of the earth, resulting in an increase in all the ocean floors.

In the 1950s, renewed interest in continental drift arose from evidence provided by the magnetic properties of fossil iron deposits. Irving (1959) and

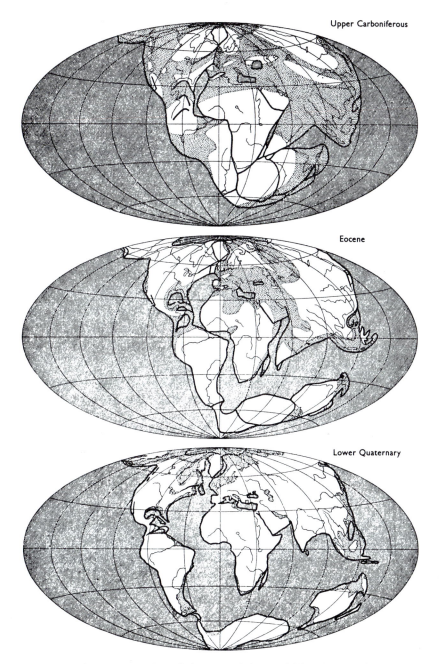

Fig. 9.15 Wegener's reconstruction of the map of the world for three geological epochs according to drift theory. (From A. Wegner, *The Origin of Continents and Oceans*, Methuen, London. Copyright © 1966, by permission of Dover Publications, Inc.)

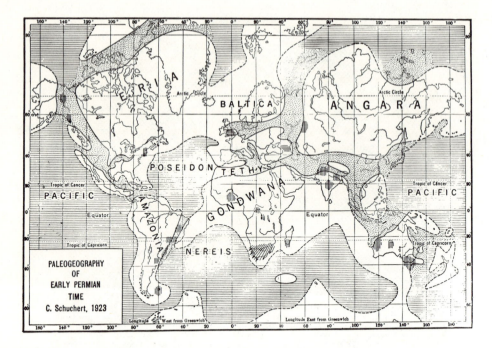

Fig. 9.16 Schuchert's reconstruction of the paleogeography and areas of glaciation in early Permian times. Oceans are ruled, epeiric seas dotted, and places of glaciation lined (vertical lines, areas of proved glaciation; horizontal lines, of uncertain glaciation). (From C. Schuchert, *A Textbook of Geology, Part II, Historical Geology*, 2nd. revised ed., John Wiley & Sons, New York. Copyright © 1924 by John Wiley & Sons, Inc., and reprinted with permission.)

Runcorn (1959) provided summary articles explaining the techniques, theory, and resultant hypotheses arising from the use of this paleomagnetic method.

The most recent theory of global dynamics, called the *plate tectonics theory*, envisages the earth's crust as a series of plates, some moving in different directions, and most of them including both continental and oceanic areas (Dietz and Holden, 1970). Theoretically, various interactions of these plates have formed deep ocean troughs, belts of volcanic activity, mountains, continental separations, and other phenomena.

As in Wegener's hypothesis, most supporters of the plate tectonics model of global geography also postulate that a united continental mass called Pangaea existed during Permian and early Triassic time, then, starting in mid-Triassic time, broke up into the present continental masses. Using radiometric ages of intrusive rocks, dated times of magnetic reversals, magnetic properties of fossil iron deposits, and other information, geophysicists have deduced dates for the times of various postulated continental separations.

Disagreement exists between geophysicists concerning a number of specific points of the plate tectonic model. For example, some believe that Africa and North America were never connected, resulting in a starting point having two supercontinents, a northern Laurasia and a southern Gondwana. Other geologists believe that before Triassic time the continental area was a single mass Pangaea, as in Fig. 15.

A summary of these ideas is given by Dietz and Holden (1970). Important steps in their model include the following:

1. In Permian and early Triassic times the continental area of the world formed a single mass stretching from pole to pole across the equator.

2. During the middle Triassic, large outpourings of basaltic rock produced rifts in this continental mass such that various segments were separated by areas of new oceanic floors.

3. By the end of Triassic time, after 20 million years of drift, (a) the northern and southern land masses were separated except at a point at Gibraltar, (b) Antarctica and Australia were separated as a separate continent, and (c) India had become detached from both Africa and Antarctica-Australia.

4. At the end of the Jurassic, after 65 million years of drift, Antarctica and Australia formed a separate southern polar continent, South America and Africa formed a unit separate from all but Laurasia, with which it was connected by a Gibraltar bridge. India was an isolated unit moving northward with unusual rapidity (geologically speaking), and North America and Eurasia formed a northern continent Laurasia, separate from the southern continents except for a connection at Gibraltar.

5. By the end of Cretaceous time, Africa and South America were well separated, Madagascar had become separate from Africa, India was passing over the equator, and Asia had come into closer contact with Africa. The North Atlantic had widened and started the separation of what is now Greenland from the central part of Laurasia. It should be noted here that Alaska and Siberia were not in close proximity, but that what is now eastern North America and western Europe were connected broadly.

6. Since Cretaceous time, the global geotectonic plates have continued to move, producing the current physiographic outlines of the earth: Notable proposed changes since the Cretaceous include; the fusion of India with Asia, the formation of a land connection between North and South America, the separation of Antarctica and Australia, a lengthening of the Atlantic producing a complete separation of Europe and North America, and

the close approximation of northeastern Asia and northwestern North America.

Assuming that the equator has always been hot and the poles cold, the generally northward movement of continents postulated by the plate tectonics theory would have brought about climatic changes on the continents as their relations to equator and poles changed. This situation would have had a profound effect on dispersals between continents. For example, according to the Dietz-Holden model, dispersals between Europe and eastern North America would have involved a warm temperate biota, whereas the much later connection between eastern Asia and western North America (when both continents were presumed to be further north) would have involved a cool temperate and/or Arctic biota.

With increasing evidence supporting the idea of an expanding earth (Meservey, 1969; Hoyle and Narlikar, 1971), there is a possibility that the Hilgenberg-Carey expanding-earth model is more probable than the Dietz-Holden model based on plate tectonics. The two models have many similarities, including the break-up of Pangaea, the separation of South America from Africa, North America from Europe, Antarctica from South America, and Australia from Africa and Antarctica. The Carey model differs in several important respects. The most important point of variance biogeographically centers around its postulates that (1) India was originally part of an African-Asian land mass and became separated from Africa by a rift causing the Arabian Sea, and (2) northeastern Asia and northwestern North America have always been connected although at times separated by a shallow Bering Sea.

It is certain that the crust of the earth has been and is dynamic, but there is no consensus among geologists concerning the patterns of its change. Considerable evidence for divergent views has been assembed by Meyerhoff (1970), Meyerhoff and Teichert (1971), and Meyerhoff and Meyerhoff (1972 *a, b*). For these reasons the systematist has little choice but to try to decide how the dispersal patterns in the group under study might have occurred according to various geotectonic theories.

Oceanic Islands

On islands that arose from the ocean floor far from any continent, such as the Galapagos Islands (about 600 miles from South America) and the Hawaiian Islands (about 2800 miles from North America), the biotic elements present must have arrived by some sort of transport. Seeds that can withstand long exposure to salt water, spores that can be carried long distances by air currents, and birds that can span long distances of ocean flight have readily established their species on these islands. But only infrequent nonvagile organisms have reached these areas and this dispersal has been primarily, if not entirely, by rafting (Darlington, 1956).

The chief mechanism of rafting seems to proceed by masses of trees, shrubs, and associated life becoming loosened from stream banks by flood, being carried out to sea by swollen river currents, and then literally being made to sail great distances over the ocean. Most of these rafts probably are either grounded on islands near their source or sink before reaching land. But at rare intervals an occasional raft has obviously reached an oceanic island and populated it with a few nonvagile species. The nonvagile biota of these islands is limited compared with continental biotas, attesting to the rarity of raft arrivals.

Intracontinental Avenues

The phylogenies of many large groups represent the occurrence of dispersals within a single continent. A knowledge of these dispersal opportunities and their times of occurrence is often helpful in arriving at a probable dating of parts of the family tree, or at least at possible dates when dispersals might have occurred.

The avenues of intracontinental dispersal are determined basically by four factors: changes in climate, mountains, epieric seas, and certain changes in genetic composition.

Climatic Changes

Climatic change alone does not result in effective dispersal from the standpoint of providing a springboard for phylogenetic branching. Let us assume, for example, that a species occupied a topographically and ecologically uniform, long, north-south land mass in the Southern Hemisphere. If the climate cooled, the species would be displaced northward; if it later warmed up, the species would be displaced southward. Such an alternation of events could go on endlessly without producing a branching of the lineage.

If the same area had sufficient ecological contrasts, then the same south-north and north-south displacements might contribute to a fission of the species and the evolution of additional lineages. To take a hypothetical example, if the climate became warmer in a land mass that was uniform ecologically except for two well-separated mountains, then those plants and animals which had adapted to cool conditions and had formerly existed in the lowlands would be displaced upward into the higher elevations of the mountains. But when this happened, the population on one mountain would be isolated from that on the other mountain if the populations were not highly vagile. Furthermore if the isolation persisted long enough, the two populations would evolve into two distinct species and another branch in the phylogeny would have occurred (see Chapter 5).

Should a climatic reversal follow these events, then each mountain species would spread into the lowland area and disperse to the base of the other mountain, resulting in a mixture of two species and setting the stage for another set of isolations if the climate again warmed up. This example applies to the biota as a whole (Ross, 1972a).

Thus changing climates must occur together with ecological or physiographic contrasts to produce intracontinental dispersals affecting speciation. The ecological and topographic contrasts are usually correlated because of the decrease in annual temperature with an increase in altitude.

Evidence from both physical geology and paleontology points to worldwide changes in temperature occurring at various times at least back to the early Paleozoic. The best documented changes are those of the last three million years, comprising the Ice Ages, or the Pleistocene. During this time, the earth experienced four major periods of extensive glaciation and reduced temperatures. In each period, world climates in even the warmest areas fluctuated in cycles having a cool phase 3–5°C (equals 6–11°F) lower than present-day temperatures (Butzer, 1964). Interglacial maxima were probably equally warmer than temperatures of the present, giving a possible difference of 6–10°C between the warm and cold pulses.

From the standpoint of the geographic dispersal of low-vagility organisms, climatic extremes are of the utmost importance. During the warm pulses, the cool-adapted species would have been restricted to the extreme northern or to the higher montane elevations preserving lower temperatures. because of latitudinal or elevational characteristics. But during the cool pulses, the cool-adapted species would have spread into the cooler lowland habitats and had the ecological opportunity to disperse widely in a manner previously impossible.

Correlated with cooler conditions of Pleistocene glacial maxima were periods of regional greater rainfall, the "pluvial periods" of Pleistocene terminology. Such pluvial periods would have provided avenues of dispersal for aquatic or mesic species of organisms between areas previously separated by tracts of hot or arid conditions in which the species could not survive. Cool, pluvial Pleistocene conditions of this type explain the dispersal of caddisflies and other aquatic insects between the Sierra Nevada of California and the Wasatch and Uinta ranges of Utah, and between the Appalachian Mountains of eastern North America and the Ozark-Ouachita mountain complex of central North America (Ross and Ricker, 1971). Similar phenomena have been inferred from the study of plants and midge faunas in New Zealand (Wardle, 1963; Craig, 1969).

During the periods of extensive cool and pluvial conditions that permitted dispersals of one type of biota, the ranges of species requiring warmer and/or zeric climates would have been reduced and many species would have been divided into isolated segments. During the next warm and drier interglacial period, the products of this latter isolation would have been able to disperse into other areas.

Evidence from intraspecific variation

Examples of dispersals so far given are based on geographic data in which related species or higher taxa have distinctive ranges. On occasion, geographic data which has been correlated with intraspecific morphological data may also indicate paths of dispersal due to climatic changes. For example, the North American winter stonefly *Allocapnia recta* occurs in a broad diagonal band from central Alabama to Maine and southeastern Canada, plus an isolated population in a peculiar artesian area in southern Alabama (Fig. 17d). In one character the species has two states, of which the ancestral one is apparent, and intermediates between them. The ancestral state is abundant in the southern isolated area where it occurs with the derived form and numerous intermediates. With the exception of a few intermediates at its extreme southern edge, the main range contains only the derived state. Because the evolution of the species occurred presumably during the Wisconsinan or last major ice age, the stages of evolution and dispersal are probably those diagrammed in Fig. 17a-d. In stage (a) the ancestral form was widespread. In a subsequent warm, dry period, the range was divided as shown in stage (b); the southern population remained ancestral, the northern one became derived. During a cooler, wetter period, stage (c), the two populations came into contact and the ancestral and derived states interbred, but only in the extreme southern area. In the most recent warming trend, the same break presumably occurred that is also postulated in stage (b), resulting in the present species distribution with a small southern isolated heterozygous population and a large northern one that is essentially homozygous for the derived state, stage (d) (Ross and Ricker, 1971).

Mountain Cycles

Geologic evidence also discloses that mountain-making events or orogenies have occurred many times in various parts of the world. Frequently, mountain chains have arisen, then been worn down or peneplaned, subsequently been reelevated, and the cycle started again. The mountains influence climate in several ways important to biotic dispersal. When they arise, they may produce corridors of cool conditions that permit the dispersal of cool-adapted species far from their ancestral home. Many long mountain ranges have a

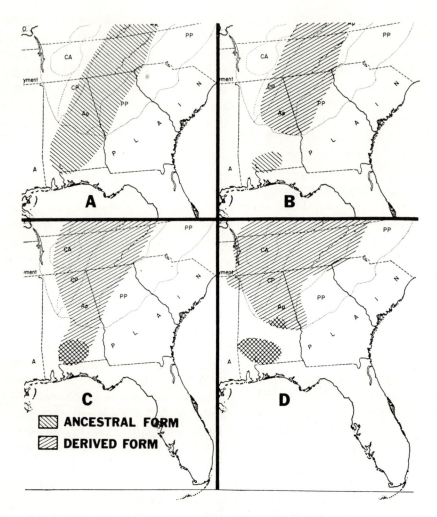

Fig. 9.17 Postulated stages in the evolution and dispersal of the ancestral and derived forms of *Allocapnia recta*. For explanation, see text. (From H. H. Ross and W. E. Ricker, "The classification, evolution, and dispersal of the winter stonefly genus *Allocapnia*," *Ill. Biol. Monog.*, **45**:1–166, 1971. Copyright © 1971 by the University of Illinois Press and reprinted with permission.)

north-south direction; these have provided dispersal routes for cool-adapted species to cross the equator. When the mountains become locally worn down and reduced in elevation, the ranges of the dispersed species will be broken up into segments, each in one of the remaining higher areas. Thus today many cool-adapted species of both plants and animals widespread in the northeastern

United States that formerly dispersed southward through the Appalachian Mountains have southern remnants persisting in the Great Smoky Mountains, a high remaining area near the southern end of the Appalachians.

If they are situated at right angles to the storm tracks of the area, mountains have a profound effect on rainfall. Because of the pressure and inversion systems produced, areas to the windward of such mountains have an unusual increase in rainfall, areas on the other side of the range have a drastically reduced rainfall producing a xeric climate. The corridor of xeric conditions on the lee side provides a corridor along which many species can disperse. The xeric conditions disrupt ranges of mesic species that formerly occupied the area. The rise of mountain ranges in mid-Cenozoic time is considered to have produced bands of xeric conditions that fragmented the formerly trans-Holarctic range of the temperate deciduous forest. If and when these ranges become reduced in elevation or, rarely, peneplaned, the more equable climates that would evolve would provide avenues for the dispersal of the biota that is now evolving in isolation on the different continents.

Epeiric Seas

During various geologic times, areas of the continental masses have been covered with extensive seas, called *epeiric seas*. Present-day examples include the South China Sea and the Euro-Asian Arctic Sea; older epeiric seas covered extensive continental areas now above water. During the Pleistocene, the alternate emergence and submergence of the area now comprising the South China Sea provided avenues of dispersal for Asiatic mainland ancestral species to reach the Philippine Islands, Borneo, Sumatra, Java, or other now isolated islands. During periods of submergence, isolated populations on each island underwent extensive speciation (Ross, 1956).

In the Paleozoic, epeiric seas were extensive over much of the Northern Hemisphere, often dividing entire continents (Schuchert, 1955). These seas may have played a major role in the evolution of terrestrial lineages.

Genetic Changes in Ecological Tolerance

It is a curious fact that, in many groups of organisms, most of the species live in one climatic zone while one or two live in quite a different zone. In the club mosses or Lycopodiaceae most of the species are tropical or subtropical, but a few occur into at least cool temperate conditions. Many examples occur in the insects. In one group of leafhoppers belonging to the genus *Empoasca,* over twenty species are tropical or subtropical, but one species occurs also in the temperate region. In the American mosquito genus *Wyeomyia,* 50 species are tropical, two are temperate. A phylogenetic analysis of the insect examples indicates that, in hit-and-miss fashion, individual lineages change genetically in a manner that allows them to disperse geographically far beyond any of their immediate relatives (Ross, Decker, and Cunningham, 1965; Ross, 1972*b*).

The genetic change confers on the species an increased ecological tolerance in some parameter. In the case of *Empoasca,* it was apparently the ability of a tropical species to live also in a frost or temperate zone. This same ability explains the northern species of club mosses and many other examples.

This frost adaptation is only one parameter of ecological change leading to geographic dispersal. Another is exemplified by the caddisfly genus *Triaenodes.* One large branch of the genus is restricted to cool, rapid streams in eastern North America, streams that are a rarity in the central part of the continent. Two species of these *Triaenodes* have evolved the ability to live also in ponds, which abound in the north-central part of the continent, and these two only have dispersed to western North America.

It seems likely that these genetic changes in ecological tolerances have played a much more important role in dispersal than has ever been suspected. If so, it will be virtually impossible to date such dispersals except when they chance to be associated with geotectonic events.

Nonvagile Aquatic Organisms

Patterns of distribution give clear testimony that aquatic organisms in continental situations have dispersed widely. Concerning vagile aquatic insects or species whose eggs or spores can be transported on the feet of aquatic birds, there is no problem. But other aquatic organisms such as fish and many aquatic crustacea have dispersed also. These do pose a problem as to how they moved from one watershed to another. It is patently impossible for them to move across an intervening strip of earth. Two mechanisms are thought to account for such interriver movements, coastal flooding and stream piracy.

Coastal Flooding

Almost all freshwater organisms can tolerate slightly brackish water for a short time. Immediately after excessive rains, rivers and streams empty such large amounts of runoff water into the coastal salt water that the latter is temporarily diluted along the shore. At the same time aquatic organisms would be swept down the water courses into coastal areas. It is believed that many freshwater species get from one drainage to the next by this combination of circumstances — being swept to sea by flood waters, then swimming into a decreasing salt gradient and at times reaching another river in the process.

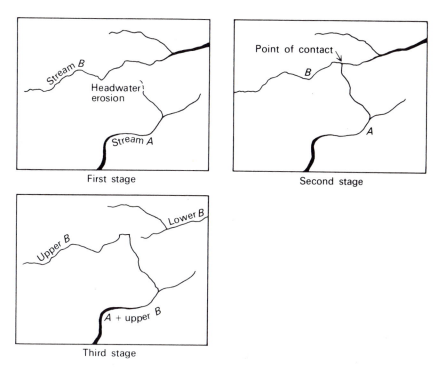

First stage

Second stage

Third stage

Fig. 9.18 Stages in the process of stream piracy.

Stream Piracy

For nonvagile aquatic species confined to headwater situations, coastal flooding would seldom, if ever, provide an intersystem dispersal pathway. For these organisms, stream piracy would seem to be the logical mechanism. Stream piracy is simple. One stream cuts back further and further into its headwater territory and eventually cuts into another stream. Under these circumstances the former would have the greater gradient and the upstream portion of the intersected stream would drain through the pirate stream (Fig. 18). Numerous examples of stream piracy have been described, the classic case (Willis, 1896) being the capture of the upper part of Beaverdam Creek by the Shenandoah River (Fig. 19).

When this happens, the headwater species inhabiting the stolen stream have the opportunity to spread throughout the pirate stream system. This mechanism is thought to be the chief one whereby headwater fishes and other species entirely restricted to the aquatic habitat spread from one region to another.

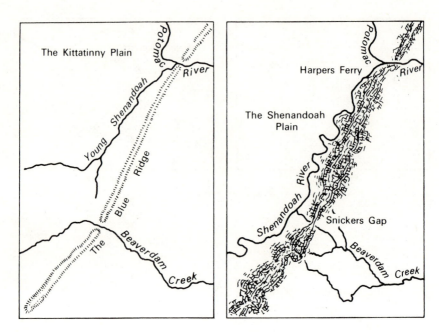

Fig. 9.19 The capture of Beaverdam Creek by the Shenandoah River, a classic example of stream piracy. Left, earlier condition; right, later condition after capture. (From B. Willis, *The Physiography of the United States,* American Book Co., New York, 1896.)

Oceanic Dispersal

The dispersal routes for oceanic life follow in many respects the same pattern as for continental life, presumably both as regards interoceanic and intraoceanic phenomena.

Interoceanic Dispersals

When the joining or separation of continents was involved, interoceanic dispersal avenues were the inverse of intercontinental avenues in that the connecting of two land areas would separate adjacent ocean areas, and vice versa. Thus the joining of North and South America by Central America connected two land masses and separated the formerly contiguous tropical waters adjacent to Central America. If part of the isthmus sank into the sea, the opposite connection and separation would occur.

It is therefore obvious that certain dispersal avenues between major oceanic areas are open to the same questions concerning moving or stable continents as are the land dispersals.

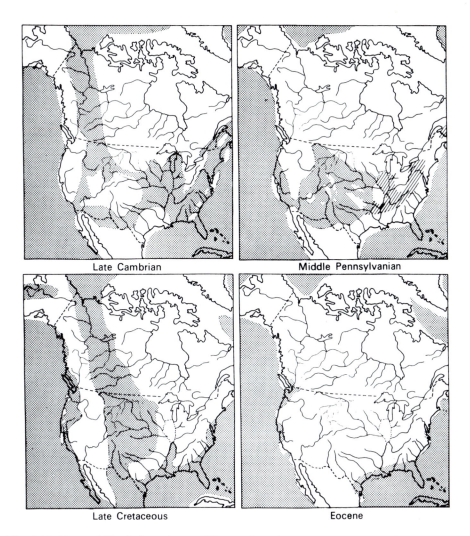

Fig. 9.20 Maps of North America at different times in the past, showing differences in the extent and location of epeiric seas. These seas were different at other times. (Adapted from Dunbar by H. H. Ross, 1963.)

Especially in the Paleozoic, epeiric seas were extensive and at times bisected entire continents (Fig. 20). Under these circumstances the epeiric seas were for the oceans what land bridges are for continents, providing an avenue for the interoceanic dispersal of all species adapted to live in any one connecting epeiric sea. At times these seas formed remarkably intricate networks by which interoceanic life became extremely widely distributed (C. A. Ross, 1967). Epeiric seas occurred also in other geologic times. Those of late Mesozoic and Cenozoic are graphically summarized and outlined by Tralau (1967).

Intraoceanic Dispersal

It is obvious from the geographic information given by Ekman (1935) that marine organisms tend to be widely distributed, as do some of the vagile continental species. On the other hand, the large number of marine genera having large numbers of species occurring in the same ocean, many with moderately restricted ranges, suggest that the oceans have mechanisms providing opportunities for the dispersal of nonvagile forms and the subsequent division of these extended ranges. The Paleozoic Fusilina studied by C. A. Ross (1967, 1969) certainly point to this conclusion.

The ecological zoning of the ocean floor necessary for such an evolutionary development was established by Petersen (1914) and extended by later authors (Moore, 1958). Whether the pelagic oceanic environments can be divided into community structure comparable to Petersen's ocean floor communities is still a moot question.

Because practically all marine species have a transportable or motile stage, it is entirely possible that in these organisms species formation and its attendant phylogenetic branching is triggered almost entirely by the mechanism of colonization. This mechanism requires no dispersal routes that owe their existence to ecological changes. The evidence is incontravertible, however, that during the climatic cycles of the Pleistocene the ocean temperatures oscillated in unison with the air temperatures. Taken in combination with the observed topographic contrasts of the ocean floor and the remarkable range of temperatures measured in various ocean currents, these temperature oscillations could have provided dispersal routes for non-vagile oceanic organisms on a par with intracontinental dispersal routes available to continental forms of life.

The exploration of these possibilities for marine organisms is one of the most challenging scientific problems confronting us.

10

Pathways of Ecological Diversification

Ecological diversity is commonplace among various taxa of organisms. In bacteria some are aerobic and some are the opposite, anaerobic. In genera scattered through plants and animals, some species are restricted to the tropics, others to temperate or even arctic conditions. In aquatic organisms, some live only in running water whereas close relatives live only in ponds or lakes.

The most striking examples of ecological diversity occur in host-specific groups such as fungi, the invertebrate parasitic phyla as the Acanthocephala, and the plant-feeding and parasitic groups of insects and mites. In each of these, species within a single genus often occur on different host species. In the fungi and plant-feeding insects it is not uncommon for a single genus to have species specific to 30 or 40 different host genera.

In spite of the wealth of examples of ecological diversity, little has been done in trying to find out the patterns by which the diversity arose. The plant genus *Vaccinium,* for example, contains both tropical and temperate species. Did the temperate species evolve from the tropical ones or vice versa? The caddisfly genus *Triaenodes* contains running-water species and still-water species. Which came first? The ironweed genus *Vernonia* contains a number of species in eastern North America, each living in a distinctive ecological setting. Which if any of these settings is the ancestral one? In instances of diverse host specificity, which was the first or ancestral host and in what order did the feeding organisms spread to other hosts? Or did the earliest members of the taxon feed on many hosts and later various species evolve a more restricted diet?

These questions lead from the static facts of ecological diversity to the dynamic question: What was the direction of ecological change? In groups exhibiting ecological diversity, which was the ancestral ecological condition and which the derived? If we had answers to these questions, it would give us great insight into the history of the ecological communities in which the species occur.

DECIPHERING ECOLOGICAL DIVERSIFICATION

When ecological diversifications are considered, several questions arise, including the following:

1. What was the direction of each?
2. How many dispersals were there in various directions?
3. At what time did each occur?
4. What was the evolutionary mechanism involved?

Direction of Diversity

Two methods have been used to decide the direction of ecological diversity — numerical and phylogenetic.

Numerical Method

It is frequently assumed that when different species occur in two opposing ecological conditions such as *tropical* vs. *temperate,* and *running water* vs. *still water,* the condition having the greater number of species was the ancestral one. The rationale of this conclusion is that the number of species now present in each condition is a function of the time the condition has been inhabited. Implicit in the explanation is the assumption that the rate of speciation would be the same in any ecological setting. Although this assumption may hold in many instances, it is far from a universal rule. Exceptions are numerous, but these have never been detected until the phylogeny of the group in question was investigated. It is therefore appropriate at this point to look at the phylogenetic method.

Phylogenetic Method

From the standpoint of the logic employed, the phylogenetic method of adducing direction of ecological diversification is exactly that used in deciphering geographical dispersal. In the latter case, first the phylogeny of the group is inferred, then the geographic data is hung on the phylogenetic tree, which becomes a sort of "hat rack." After this is done, various inferences concerning geographic dispersal may be made.

Concerning ecological diversification, the same methodology applies. After the phylogeny of the group is derived on the basis of heritable characters, then the different ecological character states can be "hung" on the "phylogenetic hat rack" and certain inferences can be made. The crux of determining direction of ecological diversification is to find out the ecological state occupied by the ancestral species (the hypothetical ancestor) of the

group, then the ecological states occupied by subsequent species. These ecological states we can call ancestral and derived.

If the group contains only two species, each occurring in a different ecological state, it is impossible to know which character state is ancestral and
which derived. Thus in Fig. 1a, if species *a* is temperate and its sister species
b is tropical, we have no way of inferring the ecological state occupied by
their ancestor 1. It could reasonably have been either tropical, subtropical,
temperate, or both tropical and temperate.

If the group contained three species with the phylogeny as in Fig. 1b,
the immediate conclusion using the fewest assumptions might be that ancestors 1 and 2 were both tropical, and that species *a* arose from tropical ancestor 2 and became temperate. If we consider possible mechanisms by which
species *a* could have become temperate, several questions arise. In the first
place, a tropical species could not become temperate while living in the tropics; therefore ancestor 2 presumably started out as a tropical species, then
evolved peripheral populations that extended into and became adapted to
temperate conditions. Possible genetic mechanisms producing this step were
outlined by Carson (1955). Subsequently, the temperate and tropical populations became geographically separated in some fashion and evolved into two
distinct species, one temperate, the other tropical. Lewis (1962) and Wallace
(1968) proposed models by which short-term but drastic climatic changes
would produce the separation. Ross et al. (1965) proposed a model based on
long-term climatic cycles as the basis of the separation. Such considerations
as these would indicate that ancestor 2 in Fig. 1b was both tropical and temperate.

If more tropical species were known that formed a longer pectinate
branch suggesting more tropical ancestors, the greater would be the probability that the tropical condition was the ancestral ecological state. Such a situation is found in seven members comprising the *Empoasca fabae* branch, a
group of American leafhoppers. The phylogeny and ecological distribution of
these species is shown in Fig. 1c, where letters have been substituted for
names in the interest of simplicity. Species *b* to *g* are either wholly tropical
or survive the winter only in frost-free areas. On this basis, ancestors 1 to 5
were almost certainly tropical. The only ancestor in doubt is 6. Using the
arguments presented concerning ancester 2 in Fig. 1b, ancestor 6 in Fig. 1c
would have been tropical plus temperate.

A comparison of ancestor 2 in Fig. 1b and ancestor 6 in Fig. 1c with
ancestor 1 in Fig. 3 of Chapter 9 indicates the essentially homologous nature
of the ecologically widespread ancestors presented here and the geographically widespread ancestors of Figs. 3 and 5 in Chapter 9. Such ancestors,

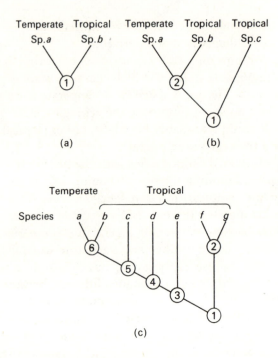

Fig. 10.1 Phylogenies involving species *a* to *g* in relation to determining direction of ecological change. For explanation, see text.

widespread either geographically or ecologically, are necessary to explain the extension of ancestral species into new areas or new ecological situations. The demonstration (or virtually so) of the logical existence of these widespread ancestors does not in itself explain the mechanisms by which either (1) the ancestral forms became widespread or (2) why one segment eventually became restricted to a new geographic or ecologic zonation. These latter phenomena involve questions of genetics and evolution rather than systematics, but questions that would not have arisen had not systematics originally brought the pertinent questions to the surface.

Examples of ecological diversity involving different hosts are extremely numerous but poorly understood. Numerous genera of fungi, parasitic protozoa, and parasitic and plant-feeding insects and mites, to name only common examples, contain dozens or hundreds of species each living on a different host, or at least many of them living on different hosts. How the host diversity arose is a moot question, although various general possibilities have been examined by Dethier (1954). The probable pathways of host diversity can

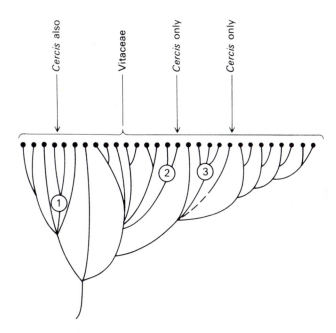

Fig. 10.2 Phylogeny and hosts of the 29 species of the *comes* species complex of the leaf-hopper genus *Erythroneura.*

best be explored by first working out the phylogeny of the consumer group (the eaters), then superimposing on this phylogenetic "hat rack" information concerning the producer taxa (the eaten). From this combination, the same type of inferences can be made as those explained above for other ecological parameters.

The *comes* species complex of the leafhopper genus *Erythroneura* offers an excellent example. Restricted almost entirely to eastern North America, the complex contains 29 species, 26 breeding only on the grape family Vitaceae, 2 only on the leguminous genus *Cercis,* and 1 on both Vitaceae and *Cercis.* It is obvious that in the evolution of the *comes* complex there has been a host transfer between the Vitaceae and *Cercis.* What was the direction? The phylogeny of the *comes* complex is shown in Fig. 2 with host data for each species. Using the same logic as outlined in parts (a) through (c) of Fig. 1, it is highly probable that the evolution of the complex occurred almost entirely on the Vitaceae, with the exception of ancestors 1, 2, and 3, each of which gave rise to a *Vitis*-inhabiting population. Those arising from ancestors 2 and 3 have evolved into distinctive species.

The host-transfer mechanism may operate in a step-by-step pattern reminiscent of the phenocline in morphology. An example is afforded by the sawfly genus *Dolerus*. The ancestral species were apparently grass-feeders, having only weak dentition of the saw, used in excavating an egg pocket in the tissues of the host grass. From these grass-feeding ancestors arose a sedge-feeding group, with the evolution of stronger dentition of the saw associated with selection pressures exerted by the problem of sawing slits and laying eggs in tougher plant tissue. From such a tough-tissue-tackling species group arose a species whose females could lay eggs in, and whose larvae could masticate the tissues of, the remarkably tough plant genus *Equisetum*. This *Equisetum*-feeding branch of *Dolerus* has evolved into a remarkably diverse and abundant branch of Holarctic phytophagous insects (Ross, 1929).

In host-specific groups the host dispersals may be extremely complex. For example, the more than 300 species of the *maculata* group of *Erythroneura*, also restricted to North America, are highly host specific and collectively breed on 17 genera of plants. The direction of dispersal would seem to be well-nigh impossible to resolve. When the host data are superimposed on the family tree of the *maculata* group, however, surprising order becomes apparent (Fig. 3a). These are simplified in Fig. 3b in which only the skeleton of the species complexes are shown, together with the hosts. The letter "Q" represents the oak genus *Quercus*. Using Occam's razor, it seems obvious that *Quercus* was the ancestral host and that many lines transferred from *Quercus* to other host genera.

In many instances, various species of a group are associated with different general ecological parameters in a community sense. Thus certain grass-feeding leafhoppers may occur in subclimax communities of a forest biome and also in similar climax communities of a prairie biome. Here the question is: Did the species or their ancestors originate in the prairie biome and later become adapted to live in a subclimax grass community of a forest biome, or vice versa?

Here again, plotting ecological characteristics in relation to phylogeny may give a probable answer to the problem. An example is the grassland leafhopper genus *Diplocolenus*. Of its six known world species, five occur only in subclimax grass communities of the sub-boreal Holarctic coniferous forest. One species occurs there and also in the climax grassland community of the central North American prairie. The question arises: Which way did evolution proceed, from species adapted to the climax grassland to those adapted to subclimax grasslands, or vice versa? When the ecological data is plotted on a probable phylogeny of the genus (Fig. 4), it appears highly probable that the primary evolution of *Diplocolenus* occurred in subclimax grass communities, and that one lineage (*D. configuratus*) became adapted to the climax prairie community (Ross and Hamilton, 1970; H. H. Ross, 1970).

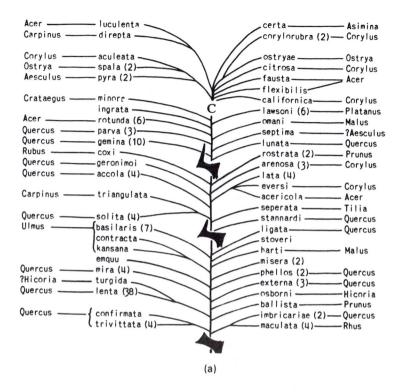

(a)

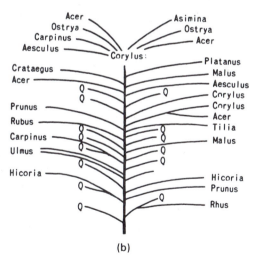

(b)

Fig. 10.3 (a) Phylogeny and known host plants of the *maculata* group of the leafhopper genus *Erythroneura*. The letter *C* indicates an inferred ancestral form occurring on *Corylus*. The black silhouettes indicate progressive changes in the style of the male genitalia. Numbers indicate the number of species in each complex. (b) The same in skeleton form with the names of the species complexes omitted and the apparent ancestral host *Quercus* indicated by the letter *Q*. (From H. H. Ross, *A Synthesis of Evolutionary Theory*, © 1962. Reprinted by permission of Prentice-Hall, Inc., Englewood Cliffs, N. J.)

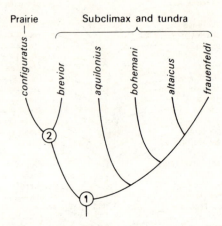

Fig. 10.4 Phylogeny and ecological affinities of the leafhoppers of the genus *Diplocolenus*. (From H. H. Ross and K. G. A. Hamilton, 1970.)

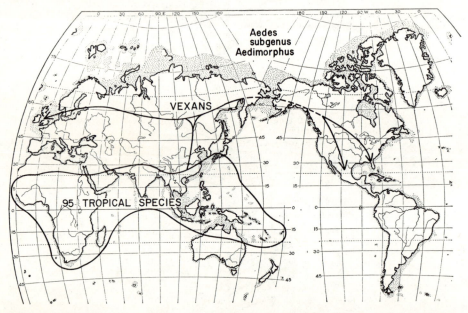

Fig. 10.5 Ecological distribution of the mosquitoes comprising the genus *Aedes* subgenus *Aedimorphus*. From H. H. Ross, "The colonization of temperate North America by mosquitoes and man," Mosquito News, 24: 103–118, 1964. Reprinted with permission.)

Number of Diversifications

If all known members of a group except one are associated with one ecological parameter, there is evidence for only one diversification along this particular pathway. For example, the mosquito subgenus *Aedimorphus* of the genus

Aedes contains 96 species. Of these, 95 are confined to the Old World tropics, but one derived species (*A. vexans*) occurs there and is also widespread across most of the Holarctic temperate region (Fig. 5). Even with only a cursory knowledge of the phylogeny, it is obvious that only one ecological diversification from tropical to temperate is involved. A similar situation occurs in the subgenus *Aedes* itself. Of 76 known species, 75 are restricted to the Asiatic and Australasian tropics, and one species occurs only in a temperate and sub-arctic band embracing the full breadth of the Holarctic region. Here again we have evidence for only one ecological dispersal between the tropical and temperate-subarctic regions.

If more than one member is associated with the second parameter, either one or more ecological diversifications may be involved. The probable number involved can again be determined by "hanging" known ecological data on the "phylogenetic hat rack" for the group. In Nearctic leafhoppers of the genus *Erythroneura,* for example, 7 species occur on sycamore (*Platanus*), 4 on deciduous holly (*Ilex*), and about 25 on trees of the apple family (Pomaceae). An integration of leafhopper phylogeny and host data indicates with a high degree of probability that (1) the 7 species on sycamore comprise a unique phylogenetic branch, hence involve only one host transfer to sycamore; (2) the four species on deciduous holly each are related to non-holly species, not to each other, involving four separate host transfers from other hosts to holly (Ross, 1953); and (3) the 25 Pomaceae species comprise one phylogenetic branch of 24 species and a distantly related branch of one species, involving two host transfers from other hosts to the Pomaceae.

Similar examples occur in the mosquitoes. In the New World branch of the tribe Sabethini, comprising about 160 species, 156 are solely tropical, 2 are tropical and subtropical, and 2 are solely temperate. How many ecological dispersals are involved? Here again, the combination of phylogeny and ecological states of the species gives the answer (Fig. 6). The two tropical-subtropical species each represent a separate ecological change. The two temperate species represent a single tropical-to-temperate diversification from an ancestral form only distantly related to the two that give rise to the two tropical-subtropical lineages.

A much different set of numerical results is afforded by certain groups of leafhoppers. In the North American *maculata* group of the leafhopper genus *Erythroneura* (Fig. 3), there is evidence of two host transfers each to *Hicoria, Carpinus, Ostrya, Prunus,* and *Ulmus;* three to *Corylus;* and four to *Acer.*

Time of Diversifications

Depending on the type of speciation process involved, ecological diversifications will have different time relations with their associated ancestors in the family tree.

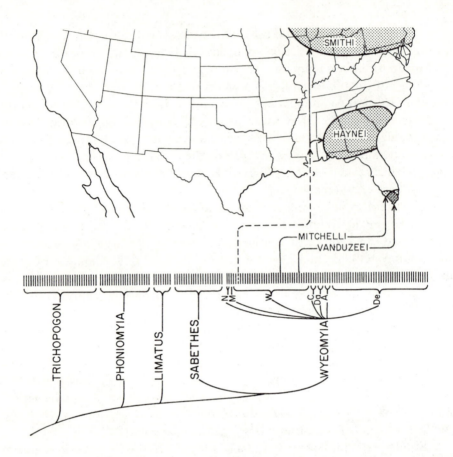

Fig. 10.6 Diagram of the phylogeny and ecological affinities of the mosquitoes comprising the New World branch of the tribe Sabethini. Each short vertical bar indicates one tropical species. (From H. H. Ross, "The colonization of temperate North America by mosquitoes and man," Mosquito News, **24:** 103–118, 1964. Reprinted with permission.)

If speciation occurred through one of the sudden processes such as polyploidy or the origin of an apomictic species, the ecological change would probably date from the origin of the new lineage. Thus in Fig. 7a, ancestor 1 occupied the ancestral ecological zone x and gave rise to a daughter population that presumably was immediately adapted to live in zone y.

If speciation occurred through processes involving gradual change of populations, such as geographic or ecological isolation, the ecological change would probably have taken place gradually (Fig. 7b). Assuming that species a appeared in derived ecological zone y and species b and c in ancestral zone x, with the phylogeny as shown (adduced from heritable characters), then

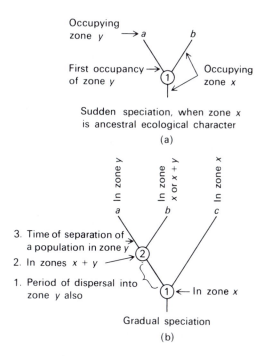

Fig. 10.7 Schematic representation of mechanisms of the evolution of ecological diversification.

ancestor 1 would have occurred in zone x and ancestor 2 would have occurred presumably in both zones. The ecological extension of the lineage into zone y could have come to pass at any time after the time of ancestor 1 but before the time of ancestor 2. Ancestor 2 would mark the time after which it broke up into sister populations, one of them restricted to zone y.

If associated with ancestral forms whose geologic age can be determined, the diversifications can be dated on this basis. Otherwise, ecological changes involved in any one phylogeny can be dated only in relation to each other — that is, which occurred first, second, and so on.

Mechanisms of Diversification and Disjunction

The mechanism of diversification would appear to require two circumstances. First, there must be some genetic or behavioral change that endows an individual and its progeny with the ability to persist in an ecological zone different from that occupied by the parental stock. Second, the change must occur in a location where it will be "selected for."

In diversifications involving climatic factors, genetic changes at the edge of the species range, coupled with normal annual climatic fluctuations, would seem to be the most likely mechanism for initiating an ecological dispersal. At the range edge, the parental type would frequently be at a disadvantage, and a mutant type better adapted to the edge conditions would be under strong selection pressure (Carson, 1955). If it became established, the edge populations would be able to disperse into areas having ecological conditions into which the parental populations had been unable to disperse.

In the many ecological characteristics that restrict the ecological distribution of aquatic species, such as tide amplitudes, current, substrate, and turbidity, there may be contrasting ecological states in spots throughout the range. An excellent example is a stream, whose flow might be slow along one stretch, fast in the next, moderate in the next, and so on. To carry this example further, streams may flow into lakes or may drain ponds, providing great ecological contrasts in close proximity. Under these circumstances, a mutant type that could persist under conditions different from those inhabited by its parents could occur almost anywhere in the species range with a considerable chance of surviving.

One of the most puzzling problems concerning ecological dispersal is the mechanism involved in host switches of host-specific species. These switches have taken place innumerable times in a great diversity of animal and plant groups. Some type of host reaction or recognition is involved, but whether it might be some nongenetic behavior perhaps akin to habituation or the result of genetic change is not known. The genetics of behavior is still too poorly understood to throw much light on this problem.

Several mechanisms may produce the disjunction of ecologically different populations in such a way as to lead to the evolution of a new species restricted to only the derived ecological zone. If the ecological change was associated with a polyploid or apomict mutation, the genetic change itself would immediately give rise to such a new species genetically isolated from its parent.

For species splitting up otherwise, some other mechanism is indicated. Lewis (1962) and Ross et al. (1965) have suggested models in which climatic changes isolate a population living entirely within the derived ecological zone, and this isolated population evolves into a separate species.

FUTURE DEVELOPMENTS

Difficulties in marshaling evidence and examples in which the direction of ecological diversification can be adduced point up the vast opportunities for research in this field. Results having a high probability of being correct require first a phylogeny carried to the species level, then reliable ecological data on these species. As correlated phylogenies and ecological data increase, we should achieve greater and greater insights into this fascinating area of scientific knowledge.

11

Classification

To communicate with each other about their findings, scientists must first give distinctive names to the objects of their study, then arrange these into some sort of orderly array. Because of the different nature of its objects of study, each major field of science has its own system of naming and arranging. In astronomy, for example, the named objects of the universe are arranged in categories ranging from atoms to clusters of galaxies. In chemistry the individually named atoms and molecules are arranged in categories ranging from small groups of similar compounds to large inclusive categories such as organic compounds and inorganic compounds. This process of naming and arranging is classification; the classification proposed for any particular set of objects is the classification of that set or group.

In practice, a classification is a descriptive arrangement not only for conversing about its included objects, but also for storing and retrieving information concerning them. Thus classification includes also devices for the identification of objects, for literature searches, and other activities helpful in the information storage-and-retrieval function.

THE HIERARCHAL SYSTEM

The classification used in biology is centered on the species, which are grouped upward into various categories such as genera and families, and are subdivided downward into lesser categories. The total of these groupings forms a set of hierarchal levels that express a nesting box system. Thus in Fig. 1, species *a* to *d* are grouped into the nesting box representing genus *A*, species *e* to *g* into the box representing genus *B*, species *h* to *l* into that of genus *C*. Genera *A* and *B*, in turn, are grouped into the nesting box representing family *A*, and genus *C* is put into the nesting box family *B*. Families *A* and *B* are grouped into the box order *A*.

It is important to note that hierarchal categories or ranks are general terms embracing all taxa at certain levels. Thus there is the hierarchal category *genus,*

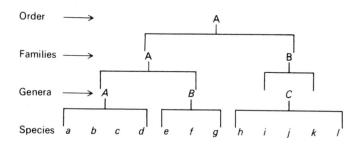

Fig. 11.1 Diagram of simple hierarchal categories and levels.

that embraces all genera; the category, *family*, that embraces all families; and
so forth. But any defined, named genus, family, order, or species is a taxon
(pl. taxa). In Fig. 1, for example, species *a* to *l* are all members of the hierarchal category *species*, but species *a* is a taxon, as are species *b, c, d, e*, etc.
Families *A* and *B* belong to the hierarchal category *family*, but family *A* is a
taxon, embracing all below it in the hierarchal system, and family *B* is also a
taxon, including all below that it embraces.

The hierarchal levels used at the time of Linnaeus included *kingdom*,
class, order, genus, species, and *variety*. Since then the major categories *phylum, division, cohort, family*, and *tribe* have been added. As the number of
known species has increased, these basic hierarchal levels above the species
became incapable of providing sufficient divisions and levels for workable
classifications, and other categories such as superfamily and subfamily, sub-
phylum, superorder and suborder are used.

The categories most in use are:

In zoology	In botany
Kingdom	Kingdom
Phylum	Division
Subphylum	Subdivision
Superclass	Class
Class	Subclass
Subclass	Order
Cohort	Suborder
Superorder	Family
Order	Subfamily

In zoology	In botany
Suborder	Tribe
Superfamily	Subtribe
Family	Genus
Subfamily	Subgenus
Tribe	Section
Genus	Series
Subgenus	Species
Species	Subspecies
Subspecies	Variety
	Form

In botany, two special categories are used for plant fossils, *organ-genus* (which is assignable to family) and *form-genus* (unassignable to family, but may be assignable to a higher taxon.

In botany and bacteriology, each species must be assigned to the ranks or categories of genus, family, order, class, and division. Other ranks are permitted if desired but not mandatory. In zoology there is no definite rule on this point but in practice each species is assigned to the basic categories genus, family, order, class, and phylum. Use of the categories other than these five basic ones is strictly up to the decision of individual systematists. Thus in groups having few representatives, such as the monotreme mammals and the horsetail plants, there is no need for more than the basic categories and only the latter are used. But in groups having extremely large numbers of species, such as the plant families Compositae and Orchidaceae, the insect families Cicadellidae and Curculionidae, and the mite order Acarina, additional categories are used. It is certain that future additions to the known biota of large taxa will require even more hierarchal levels to provide workable classifications and improvements in expressing phylogenetic relationships.

Especially among nonsystematists, there is a widespread belief that the various taxa of any hierarchal level are readily identifiable units, based on conspicuous characteristics. In certain instances they are, but in others they are not. This is not the fault of the systematist but is simply the nature of life. Many pairs or groups of species are virtually indistinguishable, yet each species is a taxon absolutely isolated genetically and evolutionarily from the others. In the higher taxa, some are indeed readily identified on the basis of differences in a single character, but others require the use of several characters for diagnosis.

An important difference exists between species and the higher categories into which they are grouped. Species are evolutionary entities whose probable limits may be determined in a scientific fashion, as outlined in Chapters 5 and 6. The assignment of species to higher categories is entirely subjective. For example, one systematist might follow the scheme of generic assignment of species shown in Fig. 1; another might place species k and l into a fourth genus D; another might place all of species a to g in genus A; and still others might propose other permutations. Each investigator would be perfectly within his rights, because there are no definite rules or tests as to what constitutes a genus, a family, or any other hierarchal level above the species.

As a result, the inclusion and definition of taxa belonging to different hierarchal levels change with time. These changes are due to a better understanding of the variational patterns of the organisms or to the discovery of additional taxa. In the 1920s, for example, the plant family Rosaceae included the roses, apples, prunes, and spireas, each typical of a separate tribe; now each of the four groups is considered a separate family by some workers. The deciduous *Rhododendrons* of eastern North America were once placed in the genus *Azalea*. Study of the genus *Rhododendron* on a worldwide basis indicated the presence of many forms intermediate between the deciduous *Azalea* species and the evergreen *Rhododendron* species. The insect family Scarabaeidae has been divided into about a dozen subfamilies; now many authors consider each of these subfamilies to be a family. Frequently changes in hierarchal rank are not based on the discovery of new characters, but solely on the opinion of the individual systematists as to which arrangement he or she thinks provides a better information storage-and-retrieval system.

NOMENCLATURE

The system of names applied to different ranks of hierarchal categories and to individual taxa constitutes the nomenclature or language of classification, and is an important part of the language of systematics as a whole. To answer the needs of science, each field requires a precise system of nomenclature, as simple as possible, and one that will be used by investigators in all countries. The ultimate aim is a system of names such that each taxon has one and only one name, and that this can be determined by workers of any land.

Like most human institutions, our present system of nomenclature is the result of a long historic process. Aristotle and the Greeks of that time had a system of classification based on logical division (Heller, 1964), but it was not sufficiently flexible to be of use in ordering the present-day known complexity of nature. The first steps in developing modern nomenclature were those leading to *binominal* nomenclature, in which the name of a species consists of two words, one the *genus,* the other the *specific name* or *epithet* (often called the *trivial* name). Prior to 1753, scientific names were *binary,* in that the distinction was made between generic and specific names; however the generic name often comprised two or three words and the specific name was often a long phrase. Occasionally each name was only one word, but in the main the names were long and cumbersome. Linnaeus adopted the binominal system of names for plants in the two volumes of *Species plantarum* (1753) and for animals in the *Systema naturae* ed. 10 (1758). These two works are the official starting points for most plant and animal systems of names. The early history of binominal nomenclature has been reviewed by Choate (1912), Stearn (1959, Heller (1964), and others.

The latinized form of scientific names is a direct outgrowth of the fact that until about the end of the 18th century most publications on natural history were written in Latin. This was the universal language of the educated people of that time. It was common for the generic name to be a Greek noun or its latinized counterpart and the specific name to be a Latin adjective, but this practice is becoming less prevalent.

The other development leading to modern nomenclature concerned the stability of names applied to taxa. For several decades after Linnaeus and his students, systematists followed Linnaeus' usage, hence stability reigned for the older names. Later there developed a trend for making names descriptive. If, for example, a species had been named *flavus* because it was yellow and someone later said it was brown, he might change the name to *brunneus.* If it were discovered that several names applied to one species or one genus, the most appropriate name might be used, or the name proposed by the most eminent of the respective authors. In these instances different investigators often chose different names for the same taxon. The situation became sufficiently chaotic that both zoological and botanical groups drafted various codes of nomenclature designed to achieve uniformity and stability in the usage of names. For plants, the first code was presented in 1813 by Augustin de Candolle in his *Theorie elementaire de la botanique*; for animals, the Strickland Code (Strickland, 1842). These and other early codes stressed *priority* as a basic tenet for choice of competing names. But priority often led to instability of names when older forgotten names were unearthed that superceded well-established names in use for many years. In later codes, various mechanisms were proposed for the retention of names based on usage through modification of the application of the law of priority. As a result, eventually international codes of nomenclature were adopted in bacteriology (1947), botany (1930), and zoology (1901), setting forth rules of conduct to be followed in the formation and use of names. Each code has been variously revised. Although each code differs in some respects from the others, each attempts to answer the same basic questions in its respective field. No code is yet available for viruses, mycoplasmae, and related organisms. Excellent brief accounts of the botanical codes are given by Lawrence (1951), of the zoological codes by Mayr (1969).

The three codes as currently revised and in effect are as follows:

International Code of Nomenclature of Bacteria, published in the *International Journal of Systematic Bacteriology* 16 (4): 459–490, October, 1966. Reprints of the English edition may be purchased from the International Microbiological Fund, 221 Science Hall, Iowa State University, Ames, Iowa 50010 U.S.A.

International Code of Botanical Nomenclature, March, 1972, published by the International Bureau for Plant Taxonomy and Nomenclature, 106 Lange Nieuwstraat, Utrecht, Netherlands.

International Code of Zoological Nomenclature, 1964, published by the International Trust for Zoological Nomenclature, 14 Belgrave Square, London, S.W. 2, England.

The principles and the legalistic aspects of the codes are discussed in this section.

Principles of Nomenclature

The basic tenets or principles of biological nomenclature are:

1. Botanical and zoological nomenclature are independent of each other. It is therefore permissible (but not desirable) for a plant genus and an animal genus to have the same name.

2. Within each nomenclature, each taxon with a particular definition, position, and rank can bear only one correct name, except in unusual or specified cases.

3. No two taxa may bear the same name.

4. Scientific names of taxa are treated as Latin names regardless of their derivation.

5. The correct name of a taxon is based upon priority of publication.

6. For the categories of order (in plants) or superfamily (in animals), and lower categories in both, the application of names of taxa is based on type specimens, type species, or type genera.

When strict application of the Rules would result in an unstable nomenclature or in ambiguity, or when the consequences of the Rules are doubtful, then in botany established custom is followed and in zoology strict application of the Code may, under specific conditions, be suspended by the International Commission on Zoological Nomenclature. In botany and bacteriology specific cases may be referred for decision to their respective International Commission or Committee.

Of paramount importance in all biological Codes is the fact that the Codes and International Commissions make judgments only on legalistic questions and problems. They have no hand in deciding the scientific interpretations of biological data. Thus, one investigator may claim that entities A and B are two distinct species, a second that they are ecotypes, and a third that they are simply ecophenotypes of one species. The Rules will make no decision on this scientific question. But whichever answer is followed, the Rules give a guide as to which name or names are to be applied to the entities under study.

BRIEF DIGEST OF THE "RULES"

The botanical and bacteriological rules are similar in general; hence if a rule is the same in both and only botany is mentioned, bacteriology is also implied.

Grammatic Form of Names

Names of taxa of suprageneric categories (tribes, families, etc.) are treated as plural substantives (nouns) in the nominative case. Names of genera, subgenera, and (in botany) sections are treated as singular substantives in the nominative case. In botany, names of certain subdivisions of genera, chiefly subsections, series, and subseries, are plurals of a contained specific epithet. For example, in the genus *Euphorbia* section *Titymalus,* the subsection *Tenellae* is based on the plural of the name of the included species epithet *E. tenella.* Such names, unlike those genera and subgenera, are treated as plural substantives. Names of species and lower taxa are adjectival, either adjectives, genitives, or nouns used in apposition. Names of genera and their formal divisions and of all suprageneric taxa begin with a capital letter. Names of species begin with a lowercase letter except in botany, where capitals are permitted in certain cases.

Limitation of the Principle of Priority

In biological nomenclature, priority does not legally extend back into antiquity but, depending on the group, is considered as starting with certain early comprehensive systematic works. The earliest starting points are various works of Linnaeus; the latest are in the 1820s. These are as follows:

Zoology

All groups start with Linnaeus, *Systema Naturae,* 10th edition, which is arbitrarily assigned the publication date of Jan. 1, 1758. Any other work published in 1758 is to be treated as after that.

Botany

Spermatophyta, Pteridophyta, Sphagnaceae, Hepaticae, Lichenes, certain algae, myxomycetes, and bacteria start with Linnaeus, *Species Plantarum,* edition 1, May, 1753. Other Musci start with Hedwig, *Species Muscorum,* Jan. 1, 1801. Fungi Caeteri start with Fries, *Systema Mycologicum,* Jan. 1, 1821. Other fungi start with Persoon, *Synopsis Methodica Fungorum,* Dec. 31, 1801. Various groups of algae have other starting dates explained in the Botanical Code. Fossil plants start with Sternberg, *Flora der Vorwelt,* Dec. 31, 1820.

Names of Taxa of Various Ranks

The principles of priority and typification do not apply to the names of taxa above the rank of family in botany and superfamily in zoology. In botany the Code makes recommendations for the endings of the names of divisions and classes, and rules for the names of orders as follows:

division	*-mycota* for fungi (Phycomycota, sac fungi) *-phyta* for others (Coniferophyta, conifers)
subdivision	*-mycotina* for fungi; *-phytina* for others
class	*-phyceae;* subclass, *-phycidae* in algae *-mycetes;* subclass, *-mycetidae* in fungi *-opsida;* subclass, *-idae* in higher plants
order	*-ales* if based on the stem of a family name (Rosales, based on *Rosa*)
suborder	*-ineae* if based on the stem of a family name (Rosineae, based on *Rosa*)

Order names such as Centrospermae, not based on the stem of a family name, follow usage.

In zoology no such rules apply, with the result that phyla, classes, orders, and suborders of animals have names bearing a variety of endings.

Family-Group Names

With certain exceptions in botany, names of families, tribes, and associated ranks are formed from the stem of an included genus, which becomes the type of the taxon, and come under the principle of priority. The category superfamily is used commonly in zoology but not in botany. Endings for the various categories are:

	Superfamily	Family	Subfamily	Tribe	Subtribe
Botany	(none)	-aceae	-oideae	-eae	-inae
Zoology	-oidea	-idae	-inae	-ini	-ina

All but endings for tribe and subtribe in zoology are specified in the Codes. These two follow traditional usage.

In botany several long-used family names not based on the stem of a generic name are treated as exceptions. Examples include the Compositae and Gramineae, which may also be called Asteraceae and Poaceae, respectively. Many recent workers in botany tend to follow the "-aceae" endings.

Generic Names

The name of a genus is a noun in the singular number, or a word treated as such. It may be taken from any source whatever, and may be an anagram or an arbitrary combination of letters. It is recommended that the name be adaptive to Latin pronunciation and not overly long. In zoology it must be written as a single word, even if compounded; in botany two words are permitted if hyphenated. For example, in zoology the hyphenated *Harlow-millsia* is invalid, but if written as the one word *Harlowmillsia*, it is valid, even though a combination of two words; in botany, the hyphenated *Uva-ursi* is valid.

Names of subgenera follow the same rules as names of genera and are considered the same rank as far as homonymy is concerned. Thus no two genera can each have a subgenus of the same name. In botany, a subgenus may be divided into sections, subsections, series, and subseries, each with a distinctive name. In both botany and zoology, the subgenus containing the type species of the genus bears the generic name unaltered; in botany the same applies to the section containing the type species of the genus. In botany, names of subsections, series, and subseries are usually the plural form of a contained species epithet, which species is implicitly the type of the taxon.

Species Names

The latinized part of the scientific name of a species comprises two words, one the generic name, the second the specific name or epithet. This epithet may be one of three types: an adjective; a genitive; or a noun in apposition, that is, a noun used as an adjective. If the specific name is an adjective, it must agree with the generic name in gender. Thus in a generic taxon having the masculine name *A- us,* a certain included species would be spelled *albus,* the masculine form of the adjective. If the species were transferred to the feminine genus *B- a*, its ending would be changed to the feminine form *alba*. Helpful hints on determining the gender of names are given by Blackwelder (1967).

If the specific name is a genitive, named after a country, collector, host, etc., its ending would not change if transferred to a genus of different gender. Thus if a species were named *smithi*, the genitive based on collector Smith, or *astraguli* based on the host *Astragulus*, it would be spelled the same if placed in the masculine genus *A- us*, the feminine genus *B- a*, or the neuter genus *C- um*. The same is true of specific names that are nouns used in apposition, such as *A- us georgia*, named after the state of Georgia; whatever the gender of genera to which it later might be transferred, its spelling would be unchanged. Each code gives rules and recommendations for the formation of genitives used as specific epithets. One notable point: in botany all genitives based on names of persons end in *-ae*. In zoology the ending of such genitives agrees with the gender and number of persons involved. Thus a genitive named for Mr. Jones would end in *-i;* for Mrs. Jones, in *-ae;* for the Jones brothers (also for Mr. and Mrs. Jones collectively), in *-orum;* and for the Jones sisters, in *-arum*.

Infraspecific Taxa

Botanical and zoological rules differ considerably on this point. In botany the infraspecific categories of subspecies, variety, form, and subform are each given latinized names. In zoology, only the subspecies is considered a valid taxonomic category nomenclaturally, and the names of subspecies are equivalent to species as regards homonymy (see later section). In botany, the rules regarding homonymy are complicated (see Botanical Code).

Formation of Names

All three codes give either rules or recommendations concerning the formation of names. In botany and bacteriology, these apply to names of divisions and below; in zoology to names of superfamilies and below. All three codes give many detailed requirements concerning the formation of names and the emendation (correction) of names originally formed improperly. These requirements necessitate a detailed knowledge of their respective Code by advanced research students in systematics.

Author Names

In botany the name of the author of a species is an official part of the specific name, plus certain other authors as specified in the botanical code. In zoology the author name is not an official part of the name but is usually given, as a bibliographic aid. The respective codes should be consulted for the appropriate rules and recommended practices concerning author citations.

Validity of Names

To become a part of the legal nomenclature of systematics, names of taxa must meet certain requirements, as follows:

1.   Publication. A name must be contained in printed matter distributed through sale, exchange, or gift to the general public or at least to institutions with libraries accessible to systematists generally. Publication is not effected by communication of new names at public meetings, or by newspaper articles, or by the placing of names in collections or exhibits open to the public, or by the issue of microfilm made from manuscripts, typescripts, or other unpublished material.

2.   The name must be accompanied by an indication, a description, an illustration, or a reference to a previously published description or illustration purportedly helpful in diagnosing the taxon. For family-group names or above, the use of a valid genus as the type of the taxon is admissible; for genus-group names, the citation of a valid species as the type-species is admissible. For species-group names the situation is more complex. In all Codes, an indication is necessary. In zoology, depending on the year of publication, a diagnostic statement and a rigid designation of the type specimen of the species may also be necessary.

Types

The name of any taxon is established by reference to a type, which may be a specimen or a taxon, depending on the category. This type is that constituent element of a taxon to which the name of the taxon is permanently attached. The purpose of the "type system" is to provide a judicial reference point for the nomenclatural identity of any taxon. The type specimen is not meant to represent all the variations in a taxon or to be typical of a taxon. There is often much confusion on the part of students and nontaxonomists on this one point.

The basic types are those designated for species and lower ranks. These are usually specimens, but in certain cases may be illustrations of specimens. When a new species is described, the entire set of specimens studied and cited by the author is considered the type series unless certain specimens are expressly excluded by the author. One specimen is designated as the *holotype,* which is the legal point of reference for the taxon; the remainder of the type series (if any) are *paratypes.* If the authors fail to designate a holotype, the entire type series are *syntypes,* and each is of equal value. One of them may be designated as the *lectotype,* which is then the legal equivalent of a holotype, and fixes the usage of the name. In botany, any specimen that is part of the single gathering (made by the collector at one time) from which the holotype was selected is an *isotype;* if the holotype is lost or destroyed, the lectotype must be selected from the existing isotypes, if there are any. If the material on which the taxon was based is missing, a specimen may be designated as a substitute type called a *neotype.* If the original type material is found later, the neotype has no further standing. The holotype, lectotype, or neotype should be given a distinctive label to facilitate its recognition and should be placed in a recognized curatorial center for preservation and availability for study. In certain groups of bacteria and protozoans the types are cultures or strains. Many of these are kept in national type culture collections.

Many workers described taxa prior to the development of the type concept. Modern authors must then study original material and designate types. Discovering the location of types is often one of the greatest problems in revisionary work. The International Association for Plant Taxonomy publishes indexes to collectors and to where workers deposited types.

The type of a genus-group name is one of its included species. If the genus-group name originally contained only one species, that one is its type. If it originally contained two or more species and if the type was not designated when the genus was named, it is designated subsequently according to various rules and recommendations in the Codes.

The type of a taxon in the family-group category is that genus from whose name the name of the family was derived. Thus the family name Rosaceae was derived from the stem of the genus *Rosa,* which is therefore the type of the family. Special cases arise in botany for family names not based on a generic name, such as the grass family Gramineae.

If its name is based on the stem of a genus, the type of a taxon in the order-group category is the included family whose name is based on the same stem. Thus the type of the order Polygonales is the family Polygonaceae; that of the order Blattaria is the family Blattidae. This practice is mandatory in bacteriology, but exists only by general implication in botany and zoology, the rules of whose codes do not extend to orders.

Homonymy

In the categories of subgenus and above, if two different taxa of the same rank are given identical names, these names are homonyms. The earlier or senior homonym is retained and the later or junior homonym must be rejected and replaced. The name of the mint genus *Tapeinanthus* Boiss. ex Benth. (1848) is a later or junior homonym of the amaryllid genus *Tapeinanthus* Herb. (1837); the former was rejected and replaced by a new name *Thuspeinata* by T. Durand in 1888.

In family-group names homonymy may result from generic names of different spelling. For example, the insect genus *Merope* and the bird genus *Merops* each resulted in the family name Meropidae. To avoid the homonymy, the International Commision on Zoological Nomenclature ruled that *Merope* should form the family name Meropeidae.

Generic and subgeneric names are coordinate in standing; hence no subgeneric name (except nominate subgenera) can be spelled exactly like any other generic or subgeneric name; similarly no new generic name can be the same as a prior subgeneric name. In botany, no two sections or lesser categories within the same genus may have the same name.

At the species level, if two different species within the same genus have the same names, these are homonyms. In this case the older name is retained and the later one is rejected and replaced. There are two kinds of homonyms, primary and secondary. If the two homonyms were originally described in the same genus, they are primary, even though the two taxa are later placed in different genera. Rejected primary homonyms may never be resurrected. If the two names were originally described in different genera, and subsequently their taxa were transferred into the same genus, they are secondary. In zoology, substitute names for junior secondary homonyms replaced after 1960 are retained only so long as the homonymy exists. Otherwise replacements of junior homonyms are permanent. For example if two species *A b-us* (1900) and *B b-us* (1920) were both transferred to genus *B,* then the later *B b-us* (1920) would be a homonym; it would be replaced by *B c-us* new name. If later the taxon *B c-us* were transferred to genus *C,* the original *B b-us* would not be in competition with *A b-us* (1900). If these were animals and *B c-us* had been named before 1961, it would be a permanent replacement for *B b-us* (1920). If *B c-us* had been named after 1960, it would be discarded and *B b-us* resurrected as the epithet of the taxon.

In zoology, whose rules recognize no categories below species and subspecies, the names of these two categories are coordinate within a genus. In botany this is not so, but it is recommended that new subspecies names should be different from species epithets in the genus.

Synonymy

Frequently two or more names are available for the same taxon. These are synonyms. The oldest name is the senior synonym, the younger ones are the junior synonyms, which latter are usually listed for each taxon with their reference and type.

For taxa in categories above order in bacteria and family or superfamily in plants and animals, respecitvely, priority does not hold and each person can use whichever name he wishes. Often certain junior synonyms become the established usage. The insect order embracing the mayflies, for example, was first called the Odontata by Latrielle in 1806, the Ephemerida by Leach in 1817, the Ephemeroptera by Haeckel in 1896, and several other names by various entomologists. The name Ephemeroptera has been almost univerally adopted at the present time. If differences of opinion exist concerning the hierarchal assignment of a taxon, the same name may be used simultaneously in two categories. For example, by some workers the insect group Homoptera is considered an order, by others the same group with the same name is considered a suborder of the order Hemiptera.

In taxa of lower categories the prior name is used, unless it is a junior homonym, in which case the next oldest name is used. On occasion the application of strict priority will upset a long-established name and lead to confusion. In such a case established usage is followed in botany, unless this is not clear; the case is then presented to the General Committee on Botanical Nomenclature, or the International Committee on Nomenclature of Bacteria. In zoology, *all* such cases must be referred to the International Committee on Zoological Nomenclature.

CONSTRUCTING A CLASSIFICATION

After a group has been studied and various taxa delineated, the next step is either to incorporate the taxa into an existing classification or to construct a new classification in whole or in part. For example, perhaps the study will reveal the existence of a number of additional species that can be fitted here and there in an existing classification without disturbing the generic or family arrangement. It might be that in another study the discovery of new characters indicates that extensive changes should be made in generic, family, and higher groupings.

Groups to be classified will be in various states of study. For some, the phylogeny may be available; for others, various statistical relationships may have been ascertained; and for others only character state tabulations may be available. In some groups the situation will be mixed. The state of knowledge may affect the type of classification that is finally resolved.

The following are the three principal bases for classifying entities at each hierarchal level into separate taxa. This would include grouping specimens or populations into species, species into genera, genera into families, and so on.

Phenetic classification

Rigid phylogenetic classification

Evolutionary classification

Phenetic Classification

In this method, groupings are based entirely on the use of characters without direct reference to phylogeny. It may be the expressed hope of the investigator that the classification constructed is natural (has phylogenetic significance), but the phylogeny as such is not expressed. Phenetic classification uses either a few selected characters, or some type of overall expression of similarity and dissimilarity using many characters.

1. *Classifying the taxa on the basis of the few characters of the group exhibiting the most conspicuous differences.* On this basis, a group of plant species having alternate leaves might be put in one genus, those with opposite leaves in another. In a group of mammals, the species with 18 teeth per jaw might be placed in one genus, those with 20 teeth per jaw in another. Formerly it was frequent for certain sets of characters to be considered especially important in making groupings at certain hierarchal levels. In the insect order Trichoptera, for example, differences in genitalia were considered the basis for sorting specimens into species, differences in wing venation for sorting species into genera, and differences in ocelli and mouthparts the basis for ordering genera into families. Similar general practices have been common in plants with regard to various flower structures, including separate or fused petals, ovary superior or inferior, number of stamens, and other structures.

This method of grouping taxa frequently results in a classification whose units can be readily identified and keyed. Its greatest drawback is that the groupings may be at variance with the natural affinities of the taxa. When these affinities are discovered, changes in the classification are usually desirable. For example, on the basis of similarities in wing venation, members of the caddisfly subfamily *Protoptilinae* were formerly placed in the family Hydroptilidae. Later, on the basis of larval characters, the *Protoptilinae* were discovered to be an offshoot of a different family, the Glossosomatidae, to which they were transferred.

2. *Classifying the taxa on the basis of overall characteristics.* The possibility that classifications based on only a few characters might produce taxa less satisfactory than those based on many characters was realized in preevolutionary times by Adanson (1757) and others. After evolution had been accepted, the same idea arose that many characters rather than a few would result in classifications more nearly expressing the path of evolution.

An ingenious set of rules for analyzing characters in this fashion was devised by Sturtevant (1939, 1942) in classifying the species of *Drosophila,* using 39 characters for the then known 58 species. It is remarkable how closely his arrangement agrees with the phylogeny of the group as it is presently understood (Throckmorton, 1968).

Since the advent of electronic computers, methods have been devised for obtaining similarity-dissimilarity data expressed in various ways, using large numbers of characters (often over 100), and for large numbers of species (Sokal and Sneath, 1963). In these studies the term *operational taxonomic units* (shortened to OTUs) is used for the entities being analyzed. If a numerical analysis is available for the group and is arranged in a phenogram or diagram, various levels of similarity may be used to designate categories in the hierarchy. In Fig. 2, for example, if the numbers represent species, they might be grouped into subgeneric units at the similarity level of 65% and into genera at level 55%. As Moss (1967) noted, subjectivity enters into the decision as to how to convert these phenograms and models into a classification. Because different types of numerical analysis of the same data may result in showing different statistical relationships between the OTUs studied, there is first a decision to be made as to which type of analysis to consider best. After this, a second decision must be made as to what levels to choose for the designation of categories. In Fig. 2, for example, the subgeneric line might be placed at the 75% line resulting in the recognition of 4 subgenera; and the generic line might be placed at 65%, resulting in 2 genera. As in other methods of classification, personal biases enter the picture such as a reticence to erect genera or subgenera on the basis of only one species, an almost subconscious leaning toward naming groups with conspicuous diagnostic characteristics and not naming others, etc.

To overcome these difficulties, Sokal and Sneath (1963) suggested the term *phenons* for groupings of OTUs based on percent similarity. As above, horizontal lines are drawn across the phenogram at various levels (Fig. 2), and entities above successive lines form a series of hierarchal categories. These may be designated as phenons possessing a certain similarity value and embracing certain entities. Thus in Fig. 2, the line at the 75% level demarks sets of four "75-phenons," embracing entities 1, 2, 5, 9; 3, 6, 7, 10; 4, and 8, respectively. The authors suggest that the phenon contents might be expressed by the first and last numbers, such as "75 phenon 1. . .9," etc. This type of nomenclature can be applied most readily and understandably when the data are presented in a two-dimensional phenogram such as Fig. 2.

Rigid Phylogenetic Classification

This includes classifications based exclusively on phylogenetic branching. Often called the *cladistic* method of classification, or *classification by recency of common descent,* it can be applied only when a phylogeny of the group is available. As first elaborated by Hennig (1950), the basic idea was that the classification of a group should also express the phylogeny precisely, render-

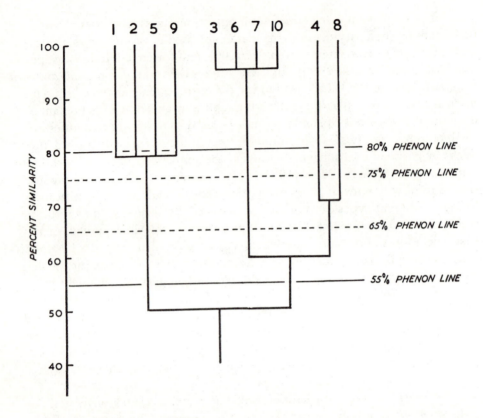

Fig. 11.2 A simple phenogram showing the formation of phenons. For explanation, see text. (From *Principles of Numerical Taxonomy* by Robert R. Sokal and Peter H. A. Sneath. W. H. Freeman and Company. Copyright © 1963. Reprinted with permission.)

ing the construction of a family tree unnecessary. Hennig's method has two requirements:

1. All taxa are monophyletic in the particular sense that each taxon includes all the species arising from a common ancestor, and no others. For example, let us suppose that the phylogeny in Fig. 3a is correct. If the entities *A, B, C,* and *D* were placed in taxon I, this would be a monophyletic grouping. The situation would fulfill the condition of the inclusion of all the species arising from the common group ancestor. If on the other hand (Fig. 3b), entities *A, B,* and *C* were placed in taxon I and entity *D* in taxon II, this relationship would no longer exist. In Hennig's sense, taxon II would be a monophyletic unit, but not taxon I. The latter would include only three of the daughter lineages of the group ancestor.

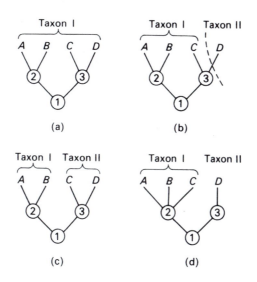

Fig. 11.3 Phylogenetic branching and classification. For explanation, see text.

2. The two sister groups of each branch are given names. Thus in Fig. 3c, taxa *A* and *B* collectively must be given a distinctive name (here called taxon I) and taxa *C* and *D* collectively must also be given a name (taxon II). Taxa I and II are each monophyletic, and each represents the total products of the two branches of ancestor 1. The resulting classification would be a precise statement of the phylogeny, thus:

> Taxon I
>> Taxon *A*
>> Taxon *B*
>
> Taxon II
>> Taxon *C*
>> Taxon *D*

If for some reason one wished to group *A*, *B*, and *C* as taxon I and to set out *D* as a separate taxon II, as in Fig. 3b, the classification would be:

> Taxon I
>> Taxon *A*
>> Taxon *B*
>> Taxon *C*
>
> Taxon II
>> Taxon *D*

On its face value, this classification would suggest the phylogeny shown in Fig. 3d, which is incorrect. If this classification were used, the phylogeny would need to be diagrammed to be apparent.

Although admittedly a cladistic classification has many good features, it also encounters objections. One is that, if rigorously applied, it normally increases considerably the number of names needed in a group. Another is that it does not take into account the amount of difference between groups, as is done in the following method.

Evolutionary Classification

In this method, which is the same as Wagner's (1969a) *groundplan/divergence* method, an effort is made to combine phylogenetic branching with the amount of evolutionary divergence (measured as character gaps) existing between different taxa. There are character gaps between all contemporaneous taxa, or we would have to classify them in some different way. But the size of the gaps differs, because of two factors. For one, lineages exhibiting intermediate character states may become extinct; more extensive extinctions produce larger character gaps between the survivors. For the other, lineages evolve at different rates; hence there will be a greater character gap between the product of a rapidly-evolving lineage and the product of a slowly-evolving sister lineage than the gap between the products of two slowly-evolving sister lineages. The combination of extensive extinction and great differences in evolutionary rates would presumably produce the largest gaps.

Why do many systematists want to include gap size in phylogenetic classifications? Because sometimes a certain sister line becomes so different character-wise from groups of other sister lines that we want to give the divergent line a special name. The reptiles, birds, and mammals (Fig. 4) make an excellent example of this point. The reptiles are a diverse group, the Recent orders including the crocodiles, lizards, snakes and turtles, and the extinct orders encompassing the dinosaurs, therapsids, and flying reptiles. From the therapsid line arose the mammals and from the crocodile line arose the birds. On strictly phylogenetic considerations, the birds would be at most a suborder of the reptile order Crocodilia, but they differ from all reptiles in so many characteristics of morphology, physiology, and especially in ecological adaptation that we want to consider the birds as an evolutionary unit comparable to reptiles as a whole. In other words, the character gap between birds and reptiles is much greater than that between the most diverse reptiles. Hence the birds are classified as the class Aves, a taxon having the same hierarchal rank as the class Reptilia from which they sprang.

In another part of the reptile tree, the mammals arose from the therapsid line. As with the birds, so the mammals differ from all existing reptiles to the

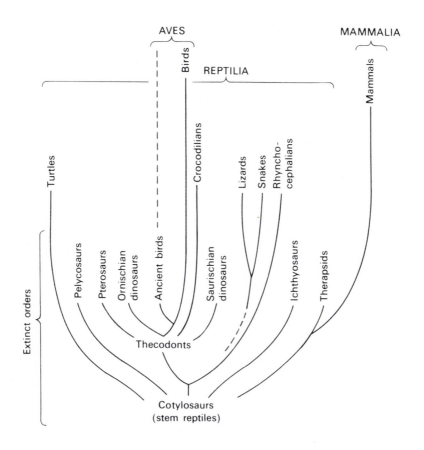

Fig. 11.4 Phylogeny and classification in the higher vertebrates. (Modified from E. H. Colbert, 1951.)

point that we consider them a separate class, Mammalia, equal in category status to the class Reptilia.

These examples and many others in both plants and animals involve taxa that were erected before evolutionary theory was developed. For this reason, authors following this practice are thought of by some as conservatives striving to retain established usage. On this score it should be pointed out that many authors follow this practice when classifying groups in which the taxa were designated recently or described as new when so placed. An example is the classification of the fruit-fly genus *Drosophila* and its allies (Fig. 5), in which about a dozen genera are recognized as arising from within the large parent genus *Drosophila.*

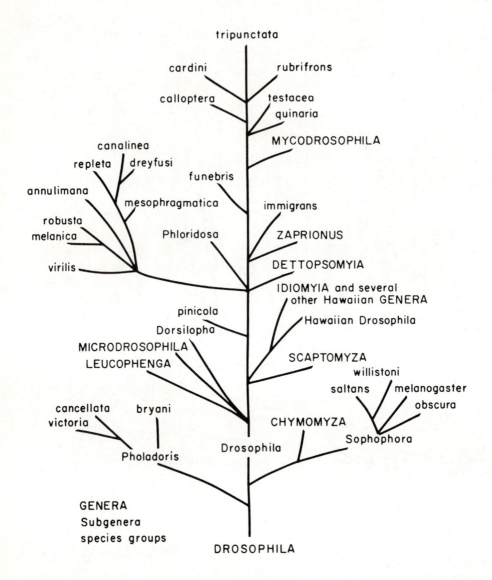

Fig. 11.5 Phylogeny of *Drosophila* and related genera. (From L. H. Throckmorton, 1962.)

MONOPHYLETIC AND RELATED TERMS

The differences in philosophy between proponents of phylogenetic (cladistic) and evolutionary classifications raise the question of the definition of *monophyletic* and its related terms, particularly *paraphyletic* and *polyphyletic*.

There have been many arguments published as to how these terms should be defined. Differences of opinion concerning these definitions have led to much ambiguity as to what particular authors meant and to arguments that were little more than exercises in semantics. It would be most helpful if an international biological committee could be appointed with the task of defining these terms rigidly as applied to systematics, but this is not possible in the foreseeable future. I will set forth here my own definitions to make my references to these terms clear.

Monophyletic

The usual dictionary definition of *monophyletic* is "of or pertaining to a single phylum; derived from or supposing animals to be derived from one parental form." Although ambiguous in certain ways, these definitions do equate *monophyletic* with the concept of arising from a common ancestor. Hennig (1950) later restricted the definition to mean that a monophyletic group included *all* the descendants of the common ancestor. But as the discussion of Ashlock (1971, 1972) pointed out, this is a special redefinition of Hennig and it does not follow general usage of the term in biological literature. As indicated by Ross (1956), all members of a monophyletic group arise from a common ancestor, but not all the lineages arising from that ancestor must necessarily be members of the designated group. An example is the genus *Drosophila* (Fig. 5). All its members arise from a common ancestor, but certain branches are excluded from the genus.

Definitions of monophyletic have usually not been concerned with the group placement of the group ancestors, and I believe that redefinitions of monophyletic and related terms with reference to these ancestors may help to clarify the concepts. I am essentially taking the broad dictionary definition as a basis and using special terms for its refinement, following the logic and nomenclature of Ashlock.

Definition of a monophyletic group

An ancestral species and all its unexcluded progeny. This definition implies two situations, one in which the group includes *all* the progeny of the ancestor and the other in which the group includes *not all* of the progeny. Ashlock has used the following names to designate these two circumstances:

Holophyletic — a group containing *all* the progeny of the original ancestor.

Paraphyletic — a group containing *not all* of the progeny of the original ancestor.

Definition of a polyphyletic group

A group whose species arise from two or more immediate ancestors, but which does not include the ultimate ancestor of these multiple immediate ancestors.

As defined above, *holophyletic* is identical with *monophyletic* as used by Hennig. As used here, *monophyletic* agrees with Simpson's (1961) opinion that the term should not necessarily reflect phylogeny but should be consistent with it. Simpson also felt that the inclusion of the ancestor in the group was theoretically sound but in practice was impractical. If, however, the phylogeny is well worked out, the hypothetical ancestor of the group (or its fossil representative if well identified) would form an integral part of the group.

The diagrams in Fig. 6 illustrate these terms in relation to the phylogenetic affinities of the component entities. In all five diagrams the known species (indicated by italic letters) grouped together in taxon X all arose from a common ancestor, and according to the above definition taxon X is monophyletic. In the diagram of part (a), taxon X includes all the progeny of the group ancestor (in this case fossil species A) and is holophyletic. In the diagrams in parts (b) and (d), taxon X does not include all the progeny of the group ancestor and by the above definitions is paraphyletic; taxon Y, containing only C, is holophyletic.

In part (d), taxon U contains species D and I, but not their ancestors. Presumably such a grouping would be based on unusual parallel evolution. As defined above, taxon U is polyphyletic. Simpson (1961) defined monophyly such that taxon U would be monophyletic. His definition (1961:124) states: "Monophyly is the derivation of a taxon through one or more lineages (temporal successions of ancestral-descendant populations) from one immediately ancestral taxon of the same or lower rank." As Hull (1964) pointed out, Simpson's definition of monophyly weakens the ability of a classification to express the sequential branchings of a phylogeny. Hennig has repeatedly pointed out that only holophyly (as defined above) can express these points, and paraphyly detracts seriously from this objective. Simpson's suggestion is another step beyond paraphyly in the weakening of a classification to express phylogeny.

If the ancestral forms are not known from fossil species but simply as hypothetical ancestors, the same reasoning can be applied. Thus part (c) illustrates the same type of relationship as does part (b), but with hypothetical ancestors substituted for fossil species. Taxon X contains not only species D, F, H, and I but also concepts of their ancestral forms back to and including ancestor 1. If species C is set out as a distinctive taxon Y, it presumably differs to an unusual derived degree from D, F, H, and I collectively and by the same token, from their inferred ancestors. Similarly, in part (e), taxon X em-

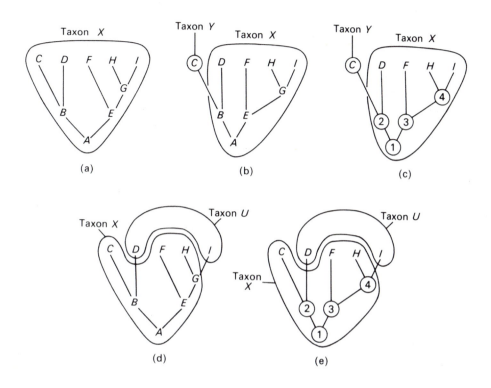

Fig. 11.6 Sample phylogenies illustrating monophyly and related terms. For explanation, see text.

braces species *C, F,* and *H* and the concepts of their hypothetical ancestors including the common ancestor 1, and is monophyletic although paraphyletic. If taxon *U* were segregated from taxon *X*, it would presumably be for the reason that species *D* and *I* differed to an unusual derived degree from *C, F,* and *H* and consequently from their hypothetical ancestors also. Taxon *U* is polyphyletic as defined here.

Most systematists make an effort to express their taxa as monophyletic groupings. Of its two kinds, holophyletic ones are usually preferable. Paraphyletic groupings are used chiefly to express situations involving the evolution of unusually distinctive branches of the group. The unusual distinctness may result from either extremely rapid evolutionary rates or from the extinction of a long series of intermediate types. In either case, later discovery of linking forms, whether represented by fossil or extant species, may add future complications to the classification.

Because (as noted above) they do not in themselves indicate the branchings of a phylogeny, when paraphyletic classifications are knowingly used the author should make the sequence of branching clear. A family tree is an excellent mechanism for conveying the desired information.

Classificatory Inflation

In 1969 Wagner discussed one of the current problems in classification: the trend toward increasing the number of taxa in each hierarchal level. He cites figures showing this trend in plants for genera, families, and orders. The same trend is common in many animal groups. Wagner further pointed out that even if the phylogeny is known, many classifications can be derived equally well from it. Using as an example a selection of representative genera of the homosporous ferns for which he had inferred the phylogeny (see Fig. 12.5, p. 306) he superimposed six classifications on the family tree (Fig. 7). The first two (a) and (b), represent earlier phenetic classifications, expressing different ideas of groupings. The last four (c) through (f), represent groupings made with a knowledge of the group's phylogeny. The classification in (c) represents one stressing dissimilarity; (d), one stressing series of branchings; (e), one stressing phenetic similarities (note one polyphyletic grouping); and (f), Wagner's preferred groupings based on phylogeny combined with phenetic considerations. As Wagner points out, it is strictly up to the opinion of the individual systematist which of many possibilities he will choose as the "best" classification for the group under study.

Previously, Edmunds (1962) proposed that family limits should be drawn such that the families could be most readily defined, would convey the greatest amount of biological information, and would be taxa whose limits provoked the least controversy. As an example, he used the classification he proposed for the higher groups of the insect order Ephemeroptera, in which there is little current argument as to the families, but great disagreement concerning the limits of superfamilies and subfamilies.

Generic Limits

The greatest divergence of views on limits of categories is found at the generic level. In animal groups having conspicuously colored or bizarre species, such as birds and many insects, there is a strong tendency to split genera into small units. At least among many entomologists, there is a trend toward defining the genus as the next recognizable taxon above the species. This leads to a situation close to that of chart (c) in Fig. 7. As the phylogeny of many of these extremely split groups becomes better known, it is likely that some of the presently recognized genera will be gathered as subgenera or lesser categories under a more inclusive genus. If not so grouped, it will be necessary

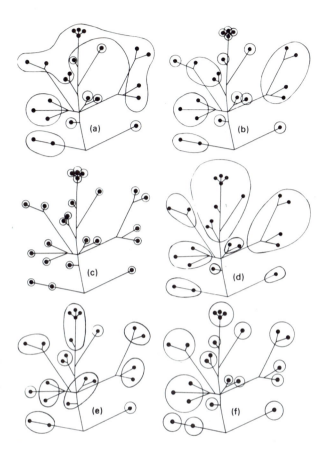

Fig. 11.7 Some possible ways of classifying the better known homosporous ferns. For explanation of letters, see text. (From W. H. Wagner, 1969.)

to erect more ranks to express any degree of relationships. Possible new categories might be supertribes (already used in aphids) and supergenera.

The first possibility (adopting wider generic limits in the future) introduces an important word of caution concerning species epithets. When describing new species in a cluster of closely related genera, be sure not to use a specific epithet that is the same as one already valid in any genus of the cluster. To do otherwise is to invite homonymy should several genera be united.

In entomology there is sentiment in some quarters for setting an upper limit (perhaps 40) to the number of species allowed in a single genus. This has led to the splitting into several genera of otherwise large and cohesive genera containing several hundred to a thousand or more species. In botany the tendency has been in the opposite direction, the use of more subdivisions of the genus (subgenus, section, etc.) and the grouping of the species into smaller taxonomic units. If the larger genus is a distinctive, readily identifiable, monophyletic taxon, such as *Carex* in the plants and *Empoasca* in the insects, this latter practice seems to me more desirable. Its great merit nomenclaturally is that it does not lead to an increase in suprageneric categories and that it preserves greater stability in nomenclature as a whole.

In various groups, efforts have been made to set specific criteria for generic separation. In the insect order Ephemeroptera, Edmunds (1962) and Peters and Edmunds (1970) recommended that species groups considered as different genera be separated by a distinct character gap in both adults and nymphs; if separated by a distinct character gap in one stage but not in both, that the groups be named as subgenera. These and recommendations of like type for other groups stress identification of all stages as a desirable feature between taxa of the same level. Often, however, the difference between distinct and indistinct character gaps is a perplexing gray zone.

Hybrid Entities

Another problem in classification is the naming of hybrid entities. The baseline of classification comprises the category of species, which are evolutionary and not arbitrary taxa as are all categories above species. The recognition of species that are of hybrid origin is not a special concern of classification because the inference that they are species is a concern of scientific investigation. Certain intergeneric hybrids, while empirical in this same fashion, pose this real problem for classification: To what genus should they be assigned? To that of one parent or the other, or to a third genus representing the original hybrid species and its progeny? Here again, the decision is one that must be made by the individual investigator (see Fig. 5.23, p. 110).

Opinions differ concerning the naming of interspecific hybrids. Four commonly used methods include:

1. naming the hybrid as a variety of the parent which it more closely resembles;

2. using a formula to indicate the parents of the hybrid, e.g., *P. alba* x *rubra;*

3. giving the hybrid a separate epithet, e.g., *P. rosea;*

4. giving the hybrid a separate epithet preceded by the multiplication sign, normally used as an indicator of hybridity, e.g., *P.* x *rosea.*

In discussing these practices, Wagner (1969*b*) stresses the importance of hybrids in contributing information to biogeography, ecology, introgression, and other areas of study. For this reason he believes that the more distinctive nonrare ones should be given names that best satisfy the criteria of convenience, stability, and meaningfulness. In his opinion the fourth method above meets these criteria best.

Interspecific hybrids are encountered commonly in many plant groups but have been noted only infrequently in animals. As a result, zoologists have named very few hybrids. On occasion, a hybrid has been described as a new species without its hybrid nature being realized, and later the name was perpetuated for the designation of hybrid individuals. For example, the two distinctive hybrid types of the warblers *Vermivora chrysoptera* and *V. pinus* are named *V. leucobronchialis* and *V. lawrencei,* respectively.

Objectives of a Classification

The subjective nature of a classification, due to the relative weight given by individual investigators to phylogeny, phenetics, time, and other criteria leads to the question: What are the objectives of a classification? It is obvious from the foregoing discussion that no consensus has been reached on this question. Hull (1964) pointed out that a classification as at present employed in biology (the Linnaean hierarchy) is a poor vehicle for conveying information in any dimension concerning phylogeny, time, rates of evolution, and others. Many of these can be expressed clearly in a properly annotated family tree. He therefore suggests that a classification be expressed as simply as possible and that the corresponding family tree be used to convey various sets of phylogenetic and evolutionary data.

This can be done readily within the limitations of monophyletic taxa if both holophyly and paraphyly are employed. The latter combines the advantages of reducing the number of necessary hierarchal categories and of expressing unusual evolutionary developments. On the other hand, polyphyletic taxa create unusual difficulties in expressing many evolutionary phenomena, and are to be avoided.

INFORMATION STORAGE AND RETRIEVAL

A fantastic amount of information concerning all known aspects of life, past and present, is stored in the scientific literature of the world. This literature now comprises millions of articles printed in books, pamphlets, sheets, journals — in fact, almost every conceivable type of publication. Such a compendium is too great for any one person to read and to digest in detail. Hence some method is needed to retrieve from the whole mass the relatively small amount of information needed at any one point of inquiry. If new information is discovered, there is needed a method to insert it into the storage memory system.

Retrieval

The initial moment of inquiry is normally when a person looks at a specimen of some organism and asks questions such as "What is its name?", "How does it breathe?", "Does it have a flower?" From the standpoint of systematics, the important first step is obtaining the name of the specimen. This is done by using successive identification aids in the following order:

1. Biological and paleontological texts. These will provide the investigator with general information concerning bacteriology, botany, zoology, and paleontology. Some of them include keys and diagnoses of the major groupings (classes or orders) of bacteria, plants, and animals. Of especial importance at this stage is learning the characters and terminology necessary to understand more detailed systematic literature.

2. Either treatises and reference books covering classes or orders, or, if available, regional or stratigraphic floras or faunas. Depending on their coverage, these will provide identification aids either to family, genus, or species.

3. Revisions of families and genera. These will give identification aids to either genus or species as they were understood at the time the revisions were made.

4. Consultation of identified museum specimens to confirm identifications made on the basis of the literature.

The four steps outlined above are preliminary. Pertinent information may have been published since the dates of the literature that was consulted. Only the week before the completion of the initial identification procedure, some investigator may have described a new species or genus, or published a more comprehensive key to some group. It might be that the only available treatise on the order was published half a century ago and since then excellent treatments had appeared on the systematics of the family. These contributions might not have been discovered in the initial identification process. It is therefore necessary to initiate a literature search for systematic papers more recent than the publications already consulted. The techniques of this literature search are different in the different fields.

In zoology, for both fossil and Recent taxa, the detailed, follow-up search begins with a perusal of the volumes of *Zoological Record*. This annual *Record* gives for each year all the publications of interest to systematists, listing all taxa described as new, all those whose status has been changed, and references to world or regional studies of particular groups. It also contains abbreviated lists of references to ecology, geographic distribution and other noneconomic topics. Each volume takes about three years in assembling and publishing.

When the most recent of these volumes have been examined, publications most likely to contain new information are also sought. But the effectiveness of this scanning depends on the library facilities available to the individual systematist. If these are not adequate, the systematist has two excellent resources available. One is to write to authors who are actively publishing in the field and to request from them either copies of, or references to, their recent publications. The other is to visit one of the systematic research centers in the area (museums, institutes, universities, state surveys, etc.) and seek advice from its staff.

In botany and bacteriology, *Biological Abstracts* (*BA*) contains references and abstracts representing a majority of the world's research literature. *Bioresearch Index* is published by the producers of *BA* and contains an almost equal number of indexed references (without abstracts) in addition to those announced in *BA*. All references from these two publications on the fungi are selected and republished in *Abstracts of Mycology;* thus, it includes much important literature for those interested in the systematic study of fungi.

In botany the most complete world listing of systematic references is the *Index Kewensis,* issued by the Kew Botanical Garden, London, England. Also helpful are the *Index to Plant Chromosome Numbers* published by the International Association for Plant Taxonomy (address on p. 263), and the *Botany Subject Index* published by the United States Department of Agriculture.

In the systematic study of fungi, termed *mycology,* the most useful general treatise is the fourth volume of *The Fungi,* by Ainsworth and Sussman (1972). References to the older taxa of fungi are contained in 25 volumes published from 1882 to 1931 by P. A. Saccardo, called *Sylloge Fungorum.* This was followed by a series of lists compiled by F. Petrak; in 1940 these were continued under the title *Index of Fungi* by the Commonwealth Mycological Institute, London, England. The same organization reprinted Petrak's *Lists.* It also publishes the *Bibliography of Systematic Mycology* and the *Review of Plant Pathology,* both useful reference sources.

For bacteria, the most definitive starting points are Bergey's *Manual* (Breed et al., 1957), giving diagnoses of all known taxa of bacteria, and Skerman's *Genera of Bacteria* (1959). These are supplemented by the *Catalogue of Cultures* published by the American Type Culture Collection, Rockville, Maryland. Colwell and Wiebe (1970) give much help with numerical problems.

Once the name of the organism has been determined, nonsystematic information may be obtained from many sources. Often the systematic literature contains references to papers giving information on distribution, ecology, life histories, and other topics. The name may be looked up in the index of publications on morphology, biochemistry, genetics, ecology, evolution, and other areas of interest, and this may lead to much desired information.

An increasing number of review journals are providing up-to-date summaries and syntheses on a wide variety of interdisciplinary topics. These contain new theoretical ideas of value in the development of systematic theory.

At present the literature search is primarily a library activity, made more efficient by learning the reference books and indices available. At least some knowledge of library science is most helpful in learning about and finding desired references. In the future, greater use will undoubtedly be made of another, faster type of literature search. Most of the major science abstracting and indexing organizations have developed or are developing systems for computerized memory banks of biological literature and techniques for retrieving from them desired sets of references by electronic machine search. Computer searches specific to the requests of a single individual are presently available in biology; their use is expected to become convention as the user becomes better informed of developments in information science and the potential these developments offer for the biologist. Biosciences Information Service (BIOSIS), the publisher of *Biological Abstracts,* also provides the faster-type, customized, individual searches mentioned above. It is developing newer systems to include in magnetic form all data needed for the production of a wide variety and flexible range of publications and search capabilities from one machine readable store, and to achieve compatibility with major systems in other fields, especially chemistry.

Storage of Information

The "store" of the storage-and-retrieval system comprises all the systematic information published in the scientific literature. To be added to this store, new information must be published in some form that is considered to constitute a legitimate part of the scientific record to which biologists and their abstracting or indexing services can refer. This excludes papers presented at meetings, newspapers, popular magazines and books, and similar outlets. Beyond that, the term scientific literature can be thought of as the types of publications that scientists read and in which they publish their scientific findings.

To ensure the effectiveness of additions to this store or data bank of systematics, the author should take some simple precautions:

1. The data should be stated as clearly and accurately as possible. This applies especially to the description of new structures and new taxa, the location of geographic points, data concerning life histories, the identification of host species, etc.

2. The data should be related to other comparable data already on record. In the case of new taxa, relations with allied taxa and diagnostic characteristics are extremely helpful.

3. If published in a journal in which systematic articles seldom appear, a copy of the article should be sent to the abstracting or indexing service most important in the field.

Keys

In systematics, analytical keys are the most effective means for the identification of taxa. Several formats are used, but the technique is simply to present a choice of characters states in each couplet, and whichever choice is made, the decision leads either to another couplet with its choices, or to the name of a taxon.

For maximum usefulness, each couplet should present clear-cut opposites in character states, such as "Flowers red" vs. "Flowers yellow." Opposites involving overlapping averages give no choice for specimens falling in the overlap zone. Thus if the opposite sides of a couplet read "Bill 7-11mm" versus "Bill 9-14mm," any specimens with a bill length of 9 to 11 mm fit either half equally well. It is argued that species are populations, not individuals, hence averages are perfectly good key characters. This argument has two flaws. First, such overlapping statements may occur in keys to supraspecific taxa, leading to utter confusion in reaching even the species-group level. Second, it is only the ability to sort individuals into discrete groups that makes possible the scientific analysis of species. Even the progeny of a single individual might represent a hybrid flock of two species, and a character-by-character analysis of every individual of the progeny would be necessary to discover this information.

Keys and Phylogeny

Some authors attempt to arrange the keys so that the successive couplets reflect the phylogeny of the group. At times the result is highly successful. Often, however, problems arise. Presumably influenced by considerations of making a more usable key, many key couplets may be based on shared ancestral characters, thus greatly weakening the portrayed phylogeny. In the opposite direction, depicting the phylogeny may necessitate the use of weak or difficult couplets, impairing ease and accuracy of identification. If either fault is evident, the two functions should be separated, the phylogeny depicted in a diagram, and the key designed for maximum accuracy of identification.

12

The Future of Systematics

The future of systematics is intimately tied in with the growing realization that we cannot speak only of the future of systematics, or only of the future of ecology, or only of the future of genetics, or only of other "futures"; we must speak of the future of biology. We have reached a stage in our investigations in which all fields are interdependent and each field is a potential contributor to the others. This relationship is certainly true of systematics, as innumerable persons have pointed out. Constance (1953) and McMillan (1954) argued for an increased communion between systematics and ecology, and Kruckeberg (1969) extended the concept to all evolutionary biology. This broad view was emphasized throughout the 1967 International Conference on Systematic Biology held at Ann Arbor, Michigan. Many of the topics that follow were discussed at greater length in the proceedings of that conference (National Research Council, 1969).

Throckmorton (1969) pointed out that the development of methods of historical analysis is probably the one function unique to systematics, and I believe that this is the key to much of its future. But this future has many facets, some representing a continuation of well-established activities and others representing an expansion of efforts that are just now attracting attention.

THE CONTINUING SEARCH

Estimates differ, but it seems certain that at least a million living species and a fantastic array of fossil species have not yet been discovered. As additional characters are found, new collecting techniques are devised, and more collections are being made in remote areas and past geologic horizons, these new additions to our knowledge of the world biota will become known. Raven (1971) has pointed out the fact that man's activities are changing the environment so extensively, to such extremes, and so rapidly that many species of organisms have already become extinct and others are doomed to follow suit.

This is especially true of local endemics that frequently hold the key to difficult phylogenetic or biogeographic problems. The dodo of Mauritius and the passenger pigeon of North America are well-known examples. Lesser publicized ones are rare species of California coastal plants exterminated by housing developments, and species of American cave insects obliterated when power dams resulted in reservoirs that flooded the caves. Raven recommends the utmost haste in collecting as many of the world's species as possible while they are still available.

An even greater effort will be needed to fill in basic knowledge concerning the species we have already described. The number known only on the basis of preserved specimens is staggering, yet we need a store of information about them that only living material can provide, including their life histories, genetics, and biochemistry.

Phylogenetic studies lag just as far behind. Only here and there in both the animal and plant kingdoms has the phylogeny of a group been investigated. Adding to our present meager coverage will require diligent searches for unrecognized characteristics and the opportunity to examine the world species of groups under study.

Computerized Systematics

One may well wonder how systematists can cope with the mass of new information being discovered through new advances in electron microscopy, electrophoresis, gas chromatography, and other developing techniques, and with the greatly enlarged collections of material envisaged in the early part of this section. One means of speeding up the handling and interpretation of the increasing number of data is the application of computer methods to steps that are normally laborious.

Use of the computer is now well established as a means of clustering taxa according to overall similarity, one of the first steps in classification and phylogeny. As mentioned on p. 186, some computer programs have been devised for inferring phylogeny. To date, the phylogenetic efforts have had only mixed success, but they do indicate that the method has great possibilities. Ingenious as man is, it is almost certain that he will be able to devise computer programs for handling successfully all the problems of phylogenetic analysis. In this effort, one of the basic difficulties that will be faced concerns the development of improved phylogenetic logic and the inference of highly probable phylogenies on which practice computer programs can be based.

Another type of computer program that has great promise as a timesaver is being developed for the identification of taxa. Dr. J. A. Peters of the U. S. National Museum devised a computer program for the identification of the genera of Central and South American snakes, and he and Dr. W. A. Weber of the University of Colorado Museum have devised another for the identification of the moss genera of Colorado. The two programs utilize a conventional data matrix of selected characters. The characters are scored for a specimen and the data are then fed into the computer. This procedure can be conducted from any location connected by teletype with the computer center; if the location is not connected to the center, the data matrix and computer program can be sent to the investigator and used on a local computer. The basic programming for this approach to computer-assisted identification was designed and developed by Mr. L. E. Morse at Michigan State University.

The North American botanists proposed a major national program of computerized data banking in systematic botany known as the Flora North America (FNA) Program, which was being directed by Mr. S. G. Shetler of the Smithsonian Institution. The aim was storing diagnostic morphological, ecological, and geographical data, along with basic nomenclatural documentation, in a generalized information retrieval system from which can be printed a synoptical Flora as well as answers to special queries as the need arises. Initially, the system was being developed to handle more or less conventional floristic/systematic data, such as traditionally appear in a Flora, and the geographic and systematic limits restrict the data base to vascular plants north of Mexico. The system is modular, however, and in principle it can be extended at any time to include other groups of organisms, other areas, and many other categories of data. The principal benefit, apart from the capability of constantly revising and updating the data bank so that current information can be made available at any time, is that automatic correlations can be made between character states and ecological or geographical parameters more or less at will. Special utility programs for mapping distributions or identifying species could be linked to the general system. Current synonymies could be produced at any time. One of the allied files which has been under development at the Smithsonian Institution for some time is the Botanical Type Specimen Register, which provides information on original publications and places of deposit of type specimens.

If these computer programs for systematics become workable and extensive, intercomputer hook-ups through telephone and desk consoles would allow systematists in all parts of the country to plug into a few large systematics data banks and extract in a short time what would take months of library scanning to accumulate. If linked with coupled closed-circuit television, the computer system could be augmented with other techniques whereby printed matter, illustrations, charts, and notes could be transmitted to the calling station and there be photographed for a permanent record.

Before such possibilities can be realized, other systematic data banks need to be developed. There are many possibilities, of which three that have already been advanced will serve as examples. Recognizing the problems imposed in systematics by poorly described species, the Entomology Section of the International Union of Biological Sciences established a central file of figures of insect types under the direction of its subsection on Figuring Type Specimens (Munroe and Beirne, 1954). Advice and cooperation were solicited from interested zoologists, but due to a lack of widespread enthusiasm the program gained little headway. In 1964 Dr. W. L. Brown of Cornell University initiated *The Pilot Register of Zoology,* a new systematics journal published in looseleaf form in which each sheet (or pair of sheets) is restricted to a single species. In 1965, through the efforts of Dr. F. A. Chace and Dr. R. S. Cowan, the Smithsonian Institution of Washington, D. C., considered the publication of a similar looseleaf journal. Due to a lack of enthusiasm and funding, the Smithsonian Institution publication never materialized and the *Register of Zoology* has been published only irregularly. Both journals were originally conceived as a means of both remedying certain faults with conventional publication practices and making possible open-ended, cumulative revisions of taxa.

The central file of figures of insect types, along with the cited two journals, offer the same excellent suggestion for making a visual computer program operable. Single unit descriptions could be put on either cards or film treated and coded for rapid magnetic machine search, so that any desired items could be called for from the requesting station, retrieved from the data bank, their images transmitted to the requestor for a photographic record, then returned to the bank. On a worldwide scale this is a gigantic undertaking, entailing not only the recording of vast amounts of old material anew, but also the resolution of many editorial, mechanical, distributional, and financial problems. But it is one possible means of speeding up systematic output.

It might appear that such highly technical developments would restrict systematic research to large institutions, that the small centers would be forced out by the cost of equipment. In many ways the opposite would be the case because a few minutes of computer time and a few hours of machine search could substitute for a specialized systematics library. By the time the systematists compile adequate data banks, electronic engineering will have improved world communication to the point that by radio telephone, line telephone, cable, or satellite transmission a systematist almost anywhere in the world could connect effectively with one of the large systematics data banks.

ENVIRONMENTS AND GENETIC EXPRESSIONS

It has long been observed that the same species may grow in a variety of habitats and that populations of the same species growing in different environments may differ in morphological traits. In many species, plants growing in bare sandy areas are dwarf, those growing in shaded areas of richer soil are luxuriant. Until the 1920s it was assumed that these growth differences were simply the expressions of a similar genome reacting with different ecological factors; in short, that they were ecophenotypes.

Working with the hawkweed *Hieracium umbellatum,* Turesson (1922, 1925) discovered that these different morphological-ecological types were also different genetically. He found similar examples in many other species. The team of Clausen, Keck, and Hiesey (Clausen, 1951) explored the same phenomenon in a variety of Californian plants. Macmillan (1959, 1964, 1965) did the same for grasses in central North America. They found genetically different ecotypes associated with soil, elevation, and climate.

A more dramatic kind of ecotype was discovered in England by Jowett (1958) – lead-tolerant races of several species of the grass genus *Agrostis* living on lead-rich soils around mine workings. Intensive investigations of these and other metal-tolerant ecotypes by British scientists (reviewed by Antonovics, 1971) has demonstrated that these ecotypes have evolved marked physiological, morphological, and genetic differences in contrast with the non-tolerant parental forms. Antedating and paralleling these studies, Kruckeberg (1954, 1969) and others discovered serpentine-tolerant races of species which were not normally thus tolerant growing on localized outcrops of serpentine soils in the western United States, and established that the serpentine tolerance was genetically controlled.

In the animal world, Dobzhansky and other students of *Drosophila* studied the genetics of natural populations and discovered a correlation between genetic structure and ecological factors in different parts of the range and/or at different seasons. In 1958 Dobzhansky reported a dramatic change in the genetic composition of southwestern populations of *Drosophila pseudoobscura* over a 17-year period, presumably due to some change in environment.

These ecological-genetic phenomena indicate that different environments impose distinctive selection pressures on the individuals growing in a specific situation and in this fashion may produce local populations having different genetic compositions. To understand a species and obtain clues concerning its evolution, it is therefore essential to know its ecological-genetic relationships over its entire range. This area of study, called *genecology,* may not seem an integral part of systematics, but it can give us insights into what is happening or what did happen in the evolution of species. It may further lead to information as to the *why* of these events. In this context, genecology may provide information as to the causal events responsible for the evolution of the taxa studied by the systematist.

Contributions to genecology are as yet meager. Informative reviews to much botanical literature have been published by Ornduff (1969) and Kruckeberg (1969), and to much zoological literature, by Mayr (1963) and Selander (1969). Tinkle (1969) pointed out some of the difficulties in obtaining comparative data on contrasting physiological and behavioral character states in natural populations that will convey reliable associated, genetic information.

Sokal and his colleagues have developed techniques for expressing and plotting character variation in relation to distribution, certainly of great importance in expressing certain types of results in genecology. Although these authors considered only morphological measurements, the differences were presumably under genetic control and subject in some fashion to natural selection pressures, i.e., selection by ecological factors. Working with aphids of the genus *Pemphigus* that make galls on *Populus,* they presented patterns of variation over the whole range (Sokal and Thomas, 1965) and patterns of microgeographic variation (Rinkel, 1965; Sokal et al., 1971).

Genecology is an interdisciplinary field of study and its results are multidirectional. In certain directions, systematics will contribute to a better understanding of ecology; in others, ecology will contribute to genetics or systematics; and so on. In spite of the small amount of illustrative material available, several facets already show promise of potential use in systematics, of which the following are the most obvious.

Identification of Physiological and Behavioral Characteristics

It is a simple matter to note differences that might be genetically controlled in biochemical or morphological traits, but often difficult to detect such differences in physiological and behavioral traits. Differences such as those found in cocoon spinning or nest construction are often conspicuous end products of behavior, but do not chronicle the differences in behavioral activities that produce the artifact. The intricasies of these activities were well demonstrated by the observations of Van der Kloot and Williams (1953*a,b*) on cocoon spinning by moth larvae. Such differences and the physiological phenomena underlying them are an integral part of the total character complement or holomorphology of the species and should be identified and compared.

By comparing lizard populations from Texas and Colorado, Tinkle found marked differences in aggressiveness, long-distance emigration, longevity, reproductive rates, and mating behavior. The two study populations belong to two subspecies of *Uta stansburiana.* If individuals from the two areas can be successfully crossed and the progeny reared, the genetic basis of their noted

differences could be determined. Considerable work of this type has been
done with plants. Klikoff (1966) isolated mitochondria from populations of
diverse habitats and found that, at a given temperature, plants of colder
regions have higher respiratory rates than those from more moderate regions.
In larvae of the midge *Chironomus tentans,* Thompson et al. (1969, 1972)
found marked differences in the hemoglobin patterns in five populations (one
from Germany and four from the United States); the genetic control has been
established, but not the environmental or functional relationships.

Mechanisms of Convergent or Parallel Evolution

In the plants, instances are now on record in which numerous species each apparently gave rise to a phenetically similar mutant type. The best examples
involve the metal-tolerant strains mentioned earlier in this section. Jowett
found that several species of grasses of the genus *Agrostis* had each produced
lead-tolerant mutants. In the western U.S. species of *Streptanthus,* Kruckeberg found serpentine-tolerant ecotypes of two normally serpentine-intolerant species and four species restricted to serpentine soils. These latter presumably all arose from one or more serpentine-intolerant parents. Phylogenetic analyses of these and other examples of parallel evolution in organisms that
can be genetically manipulated and studied ecologically should give at least
some clues on this puzzling and little understood facet of evolution. If we
had such information we should be able to devise better tests for detecting
parallelism and convergence in evolutionary studies.

Detection of Cryptic Species

One of the most intriguing questions confronting systematists is: How many
evolutionarily distinct species that we cannot detect by usual diagnosis actually
do occur today? Those we do know were recognized by comparing genetic,
behavioral, or ecological relationships between populations or individuals
formerly considered to be the same species (see p. 97). There is no theoretical reason to suppose that their mode of evolution was any different from the
evolution of sister species that can be identified readily by visible differences
in character states. Ultimately, the criterion of noninterbreeding is the clue
to their specific distinctness. Detailed studies of populations thought to belong to the same species are the best method by which to detect these cryptic
entities.

Until relatively recently, cryptic species have been discovered by inter-
population testing or random intrapopulation testing for fertility. The
recent advances in biochemical testing should provide another tool giving
clues to cryptic species. Just as Turner (1967) discovered a cryptic genus
through chemical analyses (Fig. 4), so cryptic species may be discovered
in the same fashion. Perhaps of the most use will be the various electro-

phoretic techniques developed for the detection of genetic variants of enzymes and other proteins. Smith (1960) explained many of these techniques; other authors have added to the list, including Scandalios (1969) for plants and Shaw and Prasad (1970) and Selander *et al.* (1971) for animals.

Using electrophoretic protein variants as a measure of genetic comparison between populations, two types of results are possible. One is a statistic showing differences in the amount of genetic heterozygosity between populations. Working with *Drosophila pseudoobscura*, Prakash *et al.* (1969) found that an isolated population of Bogota, Columbia, had an unusually low heterozygosity index. Breeding trials indicated that this population was probably a cryptic species. A second type of result would be detecting absolute genetic differences between populations, again suggesting breeding trials to test for cross fertility.

The number of cryptic species known at present is minuscule compared to others. This fact in no way detracts from the importance of searching for other examples. In the long run they may prove to be an extremely important part of the biota. This thought is borne out emphatically by the studies of Dobzhansky and his colleagues (Spassky et al., 1971) on the "semispecies" of *Drosophila paulistorum* (p. 79). Some biologists (Selander, 1969, and included references) appear to take the view that the persistence of cryptic species requires a set of ecological parameters different from those of readily identifiable species. I personally see no justification for this view because (1) cryptic species are merely representatives of lineages evolving independently, as are all morphologically distinct species, and (2) we know so few of the potential actual number of cryptic species that it is premature to assign them the characteristics of a mathematical ecological model. Here it is instructive to quote Ghiselin (1972): "Our universe is not populated by models, and if the organisms contradict the theories it is not the organisms that have to be corrected."

THE DIRECTION OF EVOLUTION

For many characters in various groups it is difficult if not impossible to decide from conditions within and outside the group which is the ancestral and which are the derived states. This is particularly true of characters in which various species differ in a few simple states such as *more hairy* or *less hairy,* certain parts *longer* or *shorter, darker* or *lighter,* and so on. These are the characters mentioned on p. 168 that could not be used for inferring phylogeny.

If the phylogeny has been inferred, then these indeterminate character states can be plotted on the family tree and frequently the ancestral and derived states are apparent. Kiauta (1967, 1968) used this technique to determine the ancestral and derived numbers of chromosomes in the insect orders Odonata and Trichoptera, whose chromosomes have diffuse centromeres and are therefore difficult to identify as to linkage series. In each instance, a phylogeny for the order was available. When the chromosome numbers were plotted on the respective family trees, the ancestral and derived numbers fell into cohesive patterns. In the caddisflies (Fig. 1) he inferred that chromosome evolution had proceeded chiefly by chromosome fission, with a probable ancestral number of 13. On this basis, some later fusions probably occurred in the Phryganeidae and many fusions in the Limnephilidae.

When indeterminate phenoclines are involved, the same technique is easily applied. The various states of the phenocline are plotted on the phylogenetic "hat rack" and the ancestral and derived states of the phenocline are immediately apparent. When phenoclines are composed primarily of successive loss character states, this technique frequently demonstrates that a supposed phenocline is a false one, in that similar losses have occurred independently more than once in the series. Thus an apparent phenocline may be based on taxa having 1, 2, 3, 4, or more spines on some part of the body. If the situation were as shown in Fig. 2a, it would be evident that a series of phenoclines was involved, each having lost spines independently of the others. In this example the hypothetical ancestors would exhibit a true phenocline. If the situation were as shown in Fig. 2b, the hypothetical ancestors would form a divergent pair of mirror-image phenoclines. Both sets of circumstances are frequently encountered.

Plotting indeterminate character states against phylogeny does not always result in the inference of progressive character changes. When making such comparisons for several structures of *Drosophila,* Throckmorton (1962) found evidence indicating that certain sequential ancestral forms were heterozygous for several similar states of the same character.

Behavioral Characteristics

Because of the frequent difficulty of distinguishing between genetically controlled and learned units of behavior, the direction of evolution in behavioral patterns is especially difficult to ascertain. If a probable phylogeny is available, the behavioral data may be plotted against the family tree and the direction of evolution may often be apparent. For example, in the aquatic insect order Trichoptera, larvae of certain families construct a fixed retreat, those of others construct a portable case, and still others are free living. There is considerable argument as to which was the ancestral state. Certain phenomena such as case aeration by movements of the larval abdomen were considered as supporting the view that the portable case was the ancestral state. When the phylogeny of the group was worked out, it was apparent that the

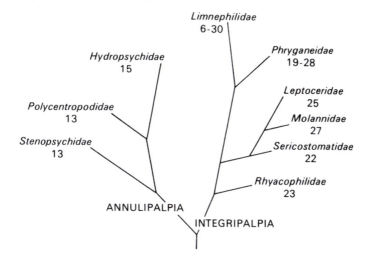

Fig. 12.1 Chromosome numbers plotted against the phylogeny of the caddisflies or Trichoptera. (Modified from B. Kiauta, 1967.)

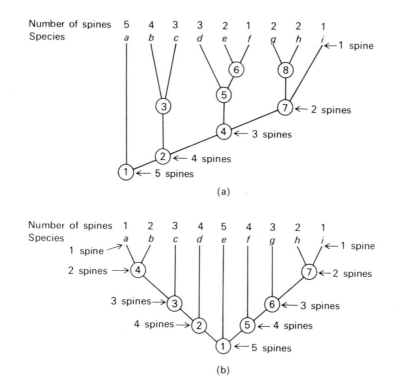

Fig. 12.2 Testing suspected phenoclines against the family tree. For explanation, see text.

fixed retreat was the ancestral state, the portable case the derived state, with a series of steps leading from one to the other (Ross, 1964).

Using the same phylogenetic technique, Selander (1964) and Selander and Mathieu (1969) found that striking differences in mating behavior in certain blister beetles had evolved between closely related species that had become sympatric. Littlejohn (1959) discovered the same situation in Australian frogs; related sympatric species had evolved different and complex mating calls, whereas allopatric relatives had simple and similar calls. In both the blister beetles and the frogs the phylogenetic evidence indicates that the simpler behavioral states are ancestral, the more complex are derived. That this behavioral progression from simple to complex is not a rule is shown by other examples. In the perching birds and the foraging bees, for example, nest-building is highly developed. Obviously, it represents a continuing and undoubtedly ancestral pattern for both groups. In both groups certain species have become parasitic on their relatives (e.g., the cow bird and cuckoo wasps) and have lost the nest-making behavior pattern. In the caddisfly case-making evolution described above, the ancestral free-living forms also lost the behavior pattern for constructing retreats.

In certain groups behavioral characteristics have proven an excellent basis for inferring the group phylogeny when morphological characteristics were inconclusive. Emerson (1938) early championed this possibility, pointing out that termite nests (representing the collective nest-building behavior of a termite colony) offered excellent material for phylogenetic inferrence. A classic application of this idea was made by Schmidt (1955a, 1955b; Emerson, 1956), who used nest characteristics as a basis for inferring the phylogeny of the termite genus *Apicotermes.*

The question as to which is the ancestral and which the derived state arises with instances of plant behavior. These states have already been inferred from the phylogeny of various groups for several examples that come to mind readily such as: evergreen or deciduous, tree or shrub or herb, wind or animal pollinated. Reversals of these trends and the direction of other behavioral phenomena will also need to be deciphered on a phylogenetic basis.

Biochemical Evolution

Interesting preliminary results indicate that the direction of biochemical evolution will prove to be a fascinating story. Chaudhuri and Chatterjee (1969), for example, discovered that in some birds L-ascorbic acid (vitamin C) is produced in the kidney; in some, in the liver; in some, in both liver and kidney; and in others, in neither. In an effort to discover some pattern in these findings, the authors plotted the various character states on a family tree of bird orders (Fig. 3). It was then evident that the ancestral enzyme systems

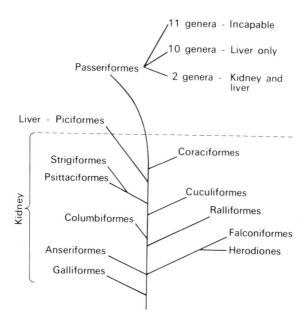

Fig. 12.3 Sites of synthesis of L-ascorbic acid in birds. (Modified from C. R. Chaudhuri and J. B. Chatterjee, 1969.)

involved occurred first in the kidney, were later (somehow) transferred to the liver, and finally, in some of the more evolved passerine birds, completely lost. The loss may have occurred more than once; a phylogeny of the passerine bird more detailed than that used here would be necessary to give information on this point. A similar trend has been observed in mammals, with the synthesis of vitamin C lost in the higher primates. This is the reason that orioles and humans need to obtain this compound from their diets.

BIOCHEMICAL SYSTEMATICS

Since about 1960 there has been an intensified interest in the use of biochemical characteristics for improving systematic conclusions. This interest is not new, but has been accelerated with the development of rapid and relatively simple techniques for chemical isolation such as paper and gas-liquid chromatography and electrophoresis. As a result, biochemical data on many classes of compounds can be assembled for large numbers of taxa. These data are providing new insights, especially into the study of hybridization, and are suggesting possibilities for phylogenetic inference. An excellent summary of the situation in plants was given by Turner (1967).

Hybridization

On the basis of morphological characters it may be difficult to be sure of the existence or the parentage of hybrids, especially hybrid swarms that may involve three or more sympatric species. In plants the simpler molecules making up the so-called secondary substances have frequently proven to be decisive in detecting hybrids and their parentage. In the genus *Baptisia,* for example, in situations involving three and four species, Alston and Turner (1963*b*) determined by paper chromatography both the hybrids and the backcrosses present, correcting conclusions based on morphological evidence. By the same technique, Smith and Levin (1963) essentially proved the parentage of the 3-way hybrid allopolyploid fern species *Asplenium kentuckiense.* In animals the larger molecules have been used for the same purpose. Manwell et al. (1967) demonstrated the hemoglobin electrophoretic patterns in midwestern sunfishes, and established criteria by which hybrid individuals could be identified with certainty. In plants, biochemical analysis has frequently demonstrated introgression which would have been difficult to detect by statistical morphological analysis. But, as Turner pointed out, some biochemical studies have given negative results when the circumstances would have indicated otherwise.

One advantage to the chromatographic method is that satisfactory extractions of at least flavonoids, alkaloids, and terpenoids can be obtained from herbarium specimens up to nearly 100 years old.

Phylogeny

Biochemical characters offer another battery of information to the phylogenist (Mabry et al., 1970). As with other characters, their usefulness is determined by how they satisfy certain criteria of uniqueness and homology, the latter concerned with the production of a certain compound by only one metabolic pathway, as discussed by Throckmorton (1968*a*).

Much information on the smaller molecules is available for plants, and it has led to intriguing discoveries. For example, in the duckweeds of the family Lemnaceae, these compounds indicate that the greatly reduced genus *Wolffia* is biphyletic (Fig. 4), one branch having arisen from a *Lemna*-like ancestor, the other from a *Wolffiella*-like ancestor (Turner, 1967). A most interesting case involves the plant order Caryophyllales or Centrospermae. In most other flowering plants the red to yellow pigments are due to anthocyanins, but in 9 of the 11 families of the Centrospermae these colors are due to betalains, an entirely different class of chemical compounds (Mabry, 1966). This circumstance will undoubtedly aid in the eventual resolution of the phylogeny within the order through a combination of information from biochemistry, embryology, and morphology (Cronquist, 1968). Bate-Smith (1972)

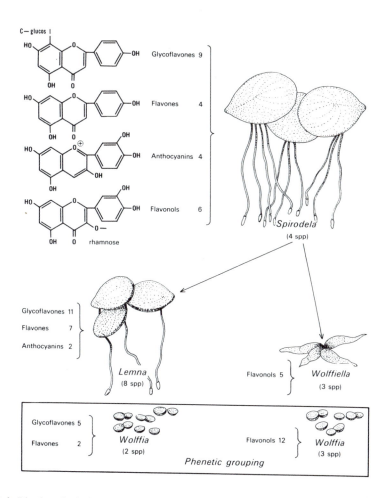

Fig. 12.4 Biochemical characters of the Lemnaceae indicating the biphyletic nature of the genus *Wolffia* as presently defined. (Illustrations revised and kindly furnished by B. L. Turner according to data in J. N. McClure and R. E. Alston (1966).)

pointed out an interesting distribution of the ellagitannins and aucubinoids in the angiosperms, and discussed several points of angiosperm phylogeny suggested by these distributions.

Few comprehensive sets of biochemical data like these are available for animals. Several sets are accumulating concerning small molecules comprising venoms and alarm substances in insects. Using the highly unique fire-ant venoms, Brand et al. (1972) were able to infer the phylogeny of a group of fire ants for which morphological data gave inconclusive results. Waterhouse and Gilby (1964), investigating the scents of bugs of the superfamily Coreoidea, and Wallbank and Waterhouse (1970), investigating defensive secretions of certain Australian cockroaches, found chemical evidence suggesting positive phylogenetic affinities for genera that had previously been placed only tentatively on the basis of morphological criteria. Moore and Wallbank (1968) tabulated defensive secretions produced by certain groups of ground beetles, the family Carabidae, and it is obvious from even a casual inspection that these data will be of great assistance in phylogenetic studies of the family.

Use of macromolecular data, both nucleic acid hybridization and comparison of amino acid sequences of homologous proteins, has as yet led only to phenetic comparisons (see p. 200). It seems certain that eventually these data can be analyzed in phylogenetic concepts. Such a development would open up a tremendous array of new information for supplementing that available from other characters. An exciting extension of the method is the fact that proteins can be extracted from certain fossils and their amino acid sequences determined. Thus Foucart et al. (1965) extracted analyzable proteins from fossil graptolites of Paleozoic time. Pauling and Zuckerkandl (1963) envisaged a future development of polypeptide phylogeny far beyond anything as yet dreamed possible.

Reviewing the admittedly scattered biochemical data bearing on systematics, it is apparent that they will form a partnership of increasing scope and value with data from other characters in systematic research.

RATES AND TYPES OF EVOLUTION

Two questions of great concern to all biologists are: How fast does evolution proceed? and, By what steps does it occur? Neither question can be answered for any group unless a highly probable phylogeny for the group has been inferred.

If the phylogeny is available, various values of evolutionary development can be superimposed on it. These normally include values for different states of individual characters, combined states of several characters of special interest, or a statistic representing changes in all observed characters. Thus one might desire to ascertain the rate at which various stages of vertebrate anatomy evolved; or how fast one set of character states evolved in relation to another set, for example, skull structure and leg structure in the emergent land tetrapods; or some measure of overall evolutionary development. Examples of, and techniques for, these studies were outlined in detail by Simpson (1944, 1953).

Wagner (1969a) devised a simple scheme that he calls the *groundplan/divergence method* that portrays graphically the amount of evolution in the group in terms of number of derived characters. Each taxon and each hypo-

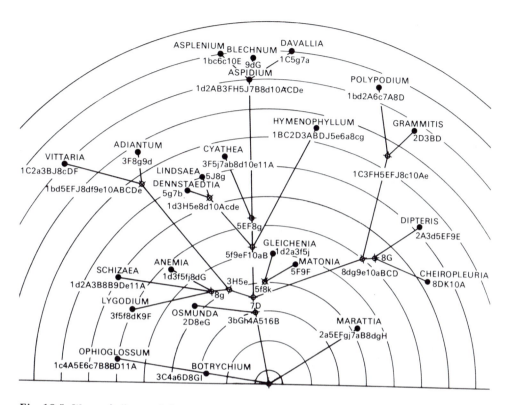

Fig. 12.5 Wagner's "groundplan/divergence" method of expressing rates of evolution as applied to homosporous ferns. For explanation, see text. (Modified from W. H. Wagner, 1969.)

thetical ancestor of the group is rated according to its total number of derived characters. The family tree is then drawn on a scaled set of concentric semicircles and the taxa and ancestors placed according to their number of derived characters. The more primitive entities and ancestors will fall in the lower central part of the chart, the more evolved ones toward the periphery (Fig. 5).

If a phylogeny is available, it may also lead to questions concerning the type of evolution involved in certain phenomena. It has long been recognized that many plant structures such as leaves or petals are under the genetic influence of "determiners" that control the shape of all leaves, or all petals, and so forth. Thus a genetic change in the leaf determiner that involved a change in leaf shape would affect all the leaves of the organism.

It is now becoming apparent through phylogenetic evidence that in many animals comparable determiners must have been influential in their evolution.

This is especially true of animals having body regions each composed of many segments, such as various parasitic worms, true worms, and arthropods. It is highly improbable that identical nephridia evolved independently in each body segment of an earthworm, or that the legs of the first five abdominal segments of the crayfish evolved independently from walking legs to identical swimming paddles. In each case it is far more reasonable to suppose the existence of a "determiner" set of genes controlling the characteristics of a set of structures. The differential evolution of the appendages of the head region of the Arthropoda is testimony that the determiner influence need not be of uniform application. But evidence from phylogeny will indicate in which groups and parts the determiner principle has been operative as an evolutionary directive.

Phylogenetic studies are leading to a reappraisal of an old question: How big are individual character changes that are important in evolution? In 1940 Goldschmidt expressed the belief that the small differences noted between species were of a different genetic nature from the large differences noted between higher taxa such as genera. The former type of change he called *microevolution,* the latter *macroevolution.* Simpson (1944) and others criticized this view, claiming that small mutational changes could account for all observed evolution, and that macroevolution was simply a result of a discontinuity in the record of change. Later, Goldschmidt (1955) expressed the thought that homeotic mutations might have played a part in macroevolution. *Homeotic mutations* are those in which the genetic control for one structure is somehow superceded by that for another. Well-known examples are the *aristipedia* and *proboscipedia* mutants in *Drosophila,* in which the apical portion of the antenna or proboscis is leg-like instead of assuming the normal shape (Fig. 6).

In reviewing the phylogeny of several insect groups, certain morphological changes suggest the operation of some such mechanism. For example, in a series of lineages of the stonefly genus *Capnia* the male has a dorsal projection on the seventh segment, used in copulation. In primitive groups of the derived daughter lineage *Allocapnia* the seventh segment has no dorsal projection but instead one used in the same manner occurs on the eighth segment. A simple explanation could be that the genetic control for this projection had moved from the seventh to the eighth segment. In some members of two advanced branches of *Allocapnia* the seventh segment also has a projection, but it is not an ancestral type and, in both cases, starts out as a smaller mime of the derived type found on the eighth segment (Fig. 7). This again suggests the operation of a type of homeotic mutation in which the genetic control of the projection on the eighth segment was added to the seventh (Ross and Ricker, 1971).

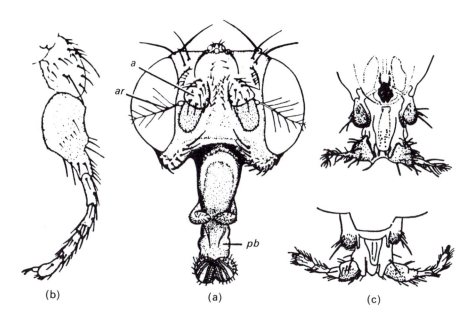

Fig. 12.6 Homeotic mutants in *Drosophila melanogaster*. (a), head with normal antennae (*a*) and proboscis (*pb*); (b) the mutant *aristipedia*; (c) two mutants of *proboscipedia*. (From C. Stern.)

A second example involves larval structures of the order Trichoptera. The ancestral types of larvae have a sclerous dorsal plate extending across the pronotum, the first segment of the thorax; the dorsum of the other two segments of the thorax is membranous. In three distantly related lines the second and third thoracic segments each have a dorsal plate that is remarkably similar to that of the first segment. These circumstances also suggest that in these three lineages some kind of homeotic mutation has caused dorsal shields to appear suddenly on the two posterior thoracic segments.

Because of the lack of credence given to Goldschmidt's ideas on homeotic mutations, little effort has been made in attempting to apply them to evolutionary studies. A second reason for this lack of interest is that a highly probable phylogeny of the study group must be available in order to know the direction of evolution of the various character states involved, and relatively few have been inferred.

SYSTEMATICS AND DECIPHERING THE EVOLUTION OF COMMUNITIES

Systematics has always had an important role in our attempts to understand the evolution of ecological communities. Phylogenetic interpretations that

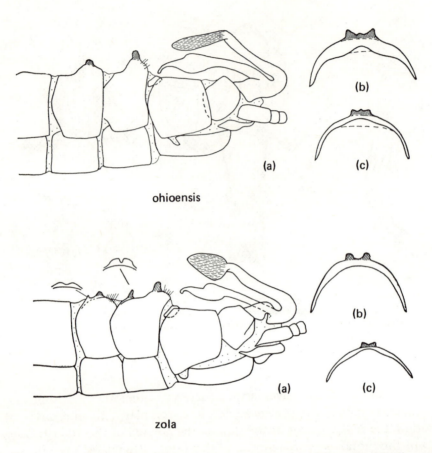

ohioensis

zola

Fig. 12.7 Dorsal projections on abdominal segments 7 and 8 of the males of *Allocapnia ohioensis* (above) and *A. zola* (below). (a), lateral aspect; (b), (c), posterior views of the projections on segments 8 and 7, respectively. (From H. H. Ross and W. E. Ricker, "The classification, evolution, and dispersal of the winter stonefly genus *Allocapnia.*" *Ill. Biol. Monog.,* **45**: 1–166, 1971. Copyright © 1971, by the University of Illinois Press and reprinted with permission.)

established the evolution of individual lineages through geologic time led to a comparison of groups of lineages that seemed to be ecologically associated. This, in turn, resulted in the formulation of a concept of communities as a whole that changed with time. As the fossil record improves, we are amassing a remarkable general picture of community evolution.

This general picture, however, does not answer questions concerning many aspects of contemporaneous ecology, which needs information on specific situations concerning community structure. Each detail of the ecological relationships in a community has a definite history, starting with the addition or deletion (extinction) of species, followed by the effect of these on interspecific reactions in food chains, competition, character displacement, coevolutionary adjustments, and so forth. The starting point with regard to species additions can eventually be answered by studies of speciation and phylogeny. The extinction facet is a difficult one that can be answered partially by neontological data but only fully by an unusually complete fossil record of the group. However in reconstructing past relationships such as competition and coevolution, it would be extremely helpful to know when and where new species were added to which communities; this includes a determination of the order (in either relative or geological time) in which various species became established in a particular community. This past is just as much a part of the community as it is of the individual lineages comprising the community; in fact, it represents *collective* pasts of these component species. It is to this past that systematics can contribute the most reliable historical element.

It is often said that we can't define a community, so why try to figure out its history. But if we had some ideas about the history of communities we would have a better basis for understanding and defining those of the present and perhaps projecting something as to their future. Whatever the skeptics say, it is my opinion that man is sufficiently curious that he will try to decipher this detailed history.

To unravel this history completely, we will need to postulate the phylogeny of all members of the past and present biotas, their geographic dispersals, and their ecological evolution. This would provide information as to when each species entered a community and as to its subsequent evolution in relation to the trophic or niche structure of that community. Undoubtedly large numbers of species now extinct played an important part in shaping present communities; much of this record we will never learn because of the incomplete nature especially of the terrestrial fossil record. Even so, that matter is not hopeless. Newer techniques for collecting fossil pollen and other small organisms, or small bits of organisms, are providing a steady increase in known biotic assemblages from many areas and times.

Much information can be gleaned from phylogenetic studies based only on living biotas. Because so few phylogenies are available, we cannot as yet delve into the evolution of entire communities. However we can obtain a great deal of information concerning parts of them. For example, plotting stream type against phylogeny in the winter stonefly genus *Allocapnia* (Ross and Ricker, 1971) indicates several points concerning the evolution of stream communities in the eastern deciduous forest of North America:

1. *Allocapnia* originally evolved in the small, rocky, cool, fairly rapid streams in the Appalachian Mountain foothills region.

2. During cooler periods of the Pleistocene, various lineages dispersed between the Appalachian region and the Ozark-Ouachita region of Missouri, Arkansas, and Oklahoma. During subsequent warm, dry periods, these species ranges were broken up and many isolated populations evolved into distinct species. About 40 of these species have persisted to the present.

3. Ancestral type streams in both areas and in connecting areas now have an *Allocapnia* fauna ranging from 2 to 8 species per stream.

4. Several species have become adapted to different types of streams (derived species in a physiological sense) and the genus has thus added species to other aquatic communities. These additions, occurring presumably at various times in the Pleistocene, include: two species to larger, faster northern streams; one species to the rapid cascades of the Great Smoky Mountains; one species to larger, warmer streams; and two species to temporary streams.

These events have altered the trophic and competitive structure in practically every running-water community of eastern North America except those of larger rivers. With some idea as to the age of these new additons, the ecologist should have in these species a powerful tool for estimating rates of ecological change produced by changes in basic diversity.

Similar situations have been discovered in certain plant groups. Through a detailed phylogenetic study of the genus *Clarkia,* for example, Lewis (1953a, b) reported that the species *C. biloba* had given rise to an unusually xeric-tolerant daughter species *C. lingulata* that has become established beyond the ecological range of other members of the genus. This means that *C. lingulata* was a biotic addition to another ecological community, with all the attendant changes in intracommunity relationships of competition and co-evolution.

Considerations of competition raise a question that has some bearing on nomenclatural practice. There is considerable reluctance, at times antagonism, among both zoologists and especially botanists to giving distinctive species names to apomict entities or autopolyploid entities that are difficult to differentiate from their parental species. Yet if the apomicts and autopolyploids form entities that are competing either with their parents or other species in the community, especially if they are causing physiological or behavioral character displacement in the competing species, then we certainly need some sort of distinctive name or "handle" to converse about them unambiguously.

Frequently one species lives in two biomes and it is not evident in which one it arose. For example, certain grass-feeding leafhoppers occur both in the climax communities of the central North American grassland biome and in subclimax grass communities of the montane forest and the boreal coniferous forest biomes to the west and north. The question arises: Did these leafhoppers evolve in the prairie biome and disperse into the others, or vice versa? In an attempt to obtain some answers to this question, the phylogeny of two leafhopper genera was inferred for the world; then the ecological distribution of each species was plotted on the family tree. In the genus *Diplocolenus* (Ross and Hamilton, 1970, Ross, 1970), all of its seven species occur in subclimax grass glades in forested areas in the northern Holarctic region; one species, *D. configuratus,* also occurs abundantly in the northern third of the North American prairie. Using Occam's razor, the evidence indicates that until recently the entire evolution of the genus took place in subclimax grass communities of the forest biomes and that relatively recently the ancestral form of *D. configuratus* evolved the ability to live in the prairie biome also. In the Holarctic genus *Rozenus,* the same situation prevails. It has many species living to the north and west of the prairie, and the derived species *R. cruciatus* is confined to the prairie itself (Ross, 1970). It is a good guess that both *D. configuratus* and *R. cruciatus* have evolved a tolerance for elevated summer temperatures, suggesting physiological tests for confirmation or denial.

In these grass-feeding leafhopper examples it would be instructive to have a phylogeny of the grasses occurring in these communities to compare with the leafhopper phylogenies. One could then determine how much the two groups had evolved in unison. Referring to the *Erythroneura* example in Fig. 3 of Chapter 10, it is obvious that the leafhoppers simply invaded one host plant genus after another long after their hosts had evolved as generic taxa.

On the other hand Seevers (1957) found evidence of concurrent evolution between termites and beetles of the family Staphylinidae that live as "guests" or inquilines in termite nests. Ahmad (1950) had previously inferred the phylogeny and dispersal of the termites. On this, Seevers superimposed the phylogeny and dispersal of the beetle guests and found a remarkable correlation (Fig. 8).

In response to selection pressures associated with life in termite nests, the various beetle lineages have evolved many peculiar glandular structures that secrete substances liked by the termites, and have also evolved many new behavior patterns associated with the utilization of these substances by the termites. The termites have evolved associated behavior patterns, but what other evolutionary changes in the termites resulted from their beetle guests is not known. A parallel situation has occurred in the case of beetles living as guests in the nests of various groups of ants. The beetles have evolved morphological and behavioral modifications as a result of the selection pressures resulting from life in ant nests, and the ants in turn have evolved at least be-

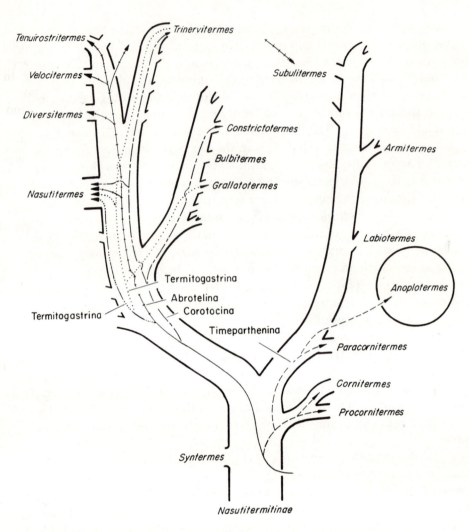

Fig. 12.8 Coevolution of termites of the subfamily Nasutitermitinae and rove beetles of the tribe Corotocini. Termite genera end in *-termes*; names ending in *-ina* are rove beetle subtribes. *Anoplotermes* is in the subfamily Amitermitinae. (From C. H. Seevers, 1957.)

havioral modifications as a result of selection pressures resulting from life with the beetles.

These evolutionary interactions between groups of organisms with close ecological relationships have been termed *coevolution* by Ehrlich and Raven (1965). Dealing primarily with the food relationships between butterfly larvae and angiosperm plants, they explain these reciprocal evolutionary relationships essentially as the occasional production by plants of chemicals that make the plant unpalatable to insects, and the occasional production by insects of mechanisms that overcome or obviate the plant "defenses." Both plant "defenses" and insect "offenses" are teleological terms, meant to express the interactions of plant and predator in terms of fortuitous genetic mutation against the background of natural selection. Ehrlich and Raven make explicit in their paper the fact that highly probable phylogenies for both plants and butterflies would shed much light on the evolutionary sequences of these events so important in understanding present ecological relationships.

Intriguing but unsolved examples of coevolution abound. One of the most fascinating concerns the fungus order Laboulbeniales and their hosts (Benjamin, 1965). Parasitic only on the animal phylum Arthropoda, the more than 100 genera and nearly 2000 known species of these tiny fungi occur chiefly on insects; a few species occur on mites (Acarina) and diplopods (Juliforma). Their insect hosts include 11 orders, over 100 families, and at least 800 genera. By far the greatest number of known Laboulbeniales occur on beetles (Coleoptera), and among these the ground beetles or Carabidae are especially "favored" hosts. When the phylogeny of both the Laboulbeniales and their hosts are known, we will have a magnificent portrayal of a complex coevolutionary progression. In relationships as complex as these it is entirely possible that a knowledge of the phylogeny of the hosts may contribute to the solution of parts of the puzzle, necessitating a truly interdisciplinary approach.

As our understanding of the evolution of communities and community relationships progresses, so will our knowledge of the taxonomic composition of previous forms of life represented in the fossil record. The two fields of study already overlap to a considerable degree, especially as regards Pleistocene events. It is certain that eventually ecological concepts based on living species will seep down further and further into the geological past and that the taxonomic findings of paleontology will improve the phylogenetic inferences on which the evolution of community diversity is based. As these two reciprocal developments expand, we will approach a vista of community evolution that at this moment seems sheer fantasy.

At this point we are left with a challenging purview of systematics. Only systematics can tell us with a high degree of probability the number and kinds of species living at any one time and in any one place. Only systematics can tell us the probable blood or phylogenetic relationships of species occurring at all times and in all places. In conjunction with geological information, systematics can give us the most probable answers as to the dispersal of lineages from one part of the world to others. In conjunction with ecological facets, systematics can give us our only reliable picture of the ecological, behavioral, and biochemical evolution of these same lineages.

Through a combination of these study areas, systematics can provide a framework for understanding how the mixtures of species we call communities came into existence, in what order the various species components entered a particular community, and (if the fossil record is adequate) how long each entity stayed there.

From this point on, evolutionary theory must take over concerning the interactions of various types of species mixtures and the fate of any one species. These interactions concern competition, mutualism, trophic interactions, and many others.

The full impact of these questions concerning community evolution — which really includes our summation of all evolution — will not be realized until systematics does two things:

1. Answers the original questions as to how many species did and do exist, what are their phylogenetic relationships, and what have been their geographic, geologic, and ecologic peregrinations.

2. Provides a classification that permits all scientists to discourse about these phenomena.

Epilogue

It is apparent that systematics is not an isolated sphere of investigation, but rather is intertwined intimately with sister fields in far-flung avenues of interdisciplinary investigation. No one field of study can any longer hope to grow alone; each contributes to and draws from the others. In this conceptual framework, systematics plays a vital and necessary role in the historical and explanatory development of the biological sciences. In this fashion it is practically thrust into the forefront of our efforts to achieve a better understanding of the universe.

Literature Cited

Anonymous, 1971. The Systematic Biology Collections of the United States: An Essential Resource. *Part I: The great collections: their nature, importance, condition and future.* The New York Botanical Garden, New York.

Abrams, L., 1905. "The theory of isolation as applied to plants." *Science*, n.s. 22:836–838.

Ahmad, M., 1950. "The phylogeny of termite genera based on imago-worker mandibles." *Bull. Amer. Mus. Nat. Hist.*, 95:37–86.

Air, G. M., E. O. P. Thompson, B. J. Richardson, and G. B. Sharman, 1971. "Amino-acid sequences of kangaroo myoglobin and haemoglobin and the date of marsupial-eutherian divergence." *Nature*, 229:391–394.

Allee, W. C., A. E. Emerson, O. Park, T. Park, and K. P. Schmidt, 1949. *Principles of Animal Ecology.* W. B. Saunders Co., Philadelphia, Pa.

Allee, W. C., and K. P. Schmidt, 1951. *Ecological Animal Geography*, 2nd. ed. John Wiley and Sons, New York.

Alston, R. E., and B. L. Turner, 1963a. *Biochemical Systematics.* Prentice-Hall, Inc., Englewood Cliffs, N. J.

——————1963b. "Natural hybridization among four species of *Baptisia* (Leguminosae)." *Amer. J. Bot.*, 50:159–173.

Ames, O., 1937. "Pollination of orchids through pseudocopulation," *Bot. Mus. Leaflets*, Harvard University 5:1–30.

Anderson, E., 1949. *Introgressive Hybridization.* John Wiley and Sons, New York.

——————, 1952. *Plants, Man and Life.* Little, Brown, Boston.

Antonovics, J., 1971. "The effects of a heterogeneous environment on the genetics of natural populations." *Amer. Scientist*, 59:593–599.

Ashlock, P. D., 1971. "Monophyly and associated terms." *Syst. Zoo.*, 20: 63–69.

——————1972. "Monophyly again." *Syst. Zool.* 21:430–438.

Axelrod, D. I., 1958. "Evolution of the Madro-Tertiary geoflora." *Bot. Rev.* 24:433–509.

Babcock, E. B., and G. L. Stebbins, Jr., 1938. "The American species of *Crepis*: Their relationships and distribution as affected by polyploidy and apomixis." *Carnegie Inst. Wash. Publ. No. 504.*

Barghoorn, E. S., and S. A. Tyler, 1965. "Microorganisms from the Gunflint chert." *Science*, 147:563–577.

Bartcher, R. L., 1966. *Fortran IV Program for Estimation of Cladistic Relationships using the IBM 7040.* Computer Contribution 6, State Geol. Surv., Univ. Kansas.

Bartholomew, J. G., W. E. Clarke, and P. H. Grimshaw, 1911. *Atlas of Zoogeography.* In Bartholomew, *Physical Atlas,* Vol. 5. Edinburgh.

Basrur, V. R., and K. H. Rothfels, 1959. "Triploidy in natural populations of the black fly *Cnephia mutata* (Malloch)." *Canad. J. Zool.,* 37:571–589.

Bate-Smith, E. C., 1972. "Chemistry and phylogeny of the Angiosperms." *Nature,* 236:353–354.

Bauer, H., 1947. "Kariologische Notizen. I. Über generative Polyploidie bei Dermapteren." *Zeits. Naturforsch,* 2:63–66.

Beçak, M. L., W. Beçak, and M. N. Rabello, 1966. "Cytological evidence of constant tetraploidy in the bisexual South American frog *Odontophrynus americanus.*" *Chromosoma,* 19:188–193.

Beckner, M., 1959. *The Biological Way of Thought.* Columbia University Press, New York.

Bell, A. W., 1959. "*Enchytraeus fragmentosus,* a new species of naturally fragmenting oligochaete worm." *Science* 129:1278.

Belozersky, A. N., and A. S. Spirin, 1960. "Chemistry of the nucleic acids of micro-organisms," pp. 147–185. In E. Chargaff and J. N. Davidson, *The Nucleic Acids.* Vol. 3, Academic Press, New York.

Benjamin, R. K., 1965. "Study in specificity." *Natural History,* 74:42–49.

Benson, R. B., 1942. "Blasticotomidae in the Miocene of Florissant, Colorado (Hymneoptera: Symphyta)." *Psyche,* 48–49:47–48.

Blackith, R. E., and R. A. Reyment, 1971. *Multivariate Morphometrics.* Academic Press, New York.

Blackwelder, R. E., 1967. *Taxonomy.* John Wiley and Sons, New York.

Blair, W. F., 1955. "Mating call and stage of speciation in the *Microhyla olivacea–M. carolinensis* complex." *Evolution,* 9:469–480.

Blair, W. F., and W. E. Howard, 1944. "Experimental evidence of sexual isolation between three forms of mice of the cenospecies *Peromyscus maniculatus.*" *Contrib. Lab. Vert. Biol. Univ. Mich.,* 26:1–19.

Bohart, R. M., 1941. "A revision of the Strepsiptera with special reference to the species of North America." *Univ. Calif. Pub. Ent.,* 7 (6):91–160.

Brand, J. M., M. S. Blum, and H. H. Ross, 1973. "Biochemical evolution in fire ant venoms." *Insect Biochemistry,* 3: 45–51.

Breed, R. S., E. G. D. Murray, N. R. Smith, and others, 1957. Bergey's *Manual of Determinative Bacteriology.* Williams and Wilkins Co., Baltimore, Md.

Briggs, D., and S. M. Walters, 1969. *Plant Variation and Evolution.* McGraw-Hill, World Univ. Library, New York.

Brink, R. A., 1960. "Paramutation and chromosome organization." *Quart. Rev. Biol.,* 35:120–137.

————1964. "Genetic repression in multicellular organisms." *Amer. Nat.,* 98:193–211.

Broecker, W. S., 1965. "Isotope geochemistry and the Pleistocene climatic record." In *The Quaternary of the United States,* H. E. Wright, Jr., and D. G. Frey, eds., Princeton Univ. Press, New Jersey. pp. 737–753.

Buchanan, L. L., 1939. "The species of *Pantomorus* of America north of Mexico." *U. S. D. A. Misc. Publ.*, 341:1–39.

Bush, G. L., 1969. "Sympatric host race formation and speciation in frugivorous flies of the genus *Rhagoletis* (Diptera, Tephritidae)." *Evolution*, 23:237–251.

Butts, R. E., 1969. *William Whewell's Theory of Scientific Method.* University of Pittsburgh Press, Pittsburgh.

Butzer, K. W., 1964. *Environment and Archeology.* Aldine Publ. Company, Chicago.

Cain, S. A., 1944. *Foundations of Plant Geography.* Harper and Bros., New York.

Camin, J. H., and R. R. Sokal, 1965. "A method for deducing branching sequences in phylogeny." *Evolution*, 19:311–326.

Camp, H. L., and C. L. Gilly, 1943. "The structure and origin of species." *Brittonia*, 4:323–385.

Carey, S. W., 1958. "The tectonic approach to continental drift." In *Continental Drift: a Symposium*, pp. 177–355. Geology Department, University of Tasmania, Hobart.

_____1970. "Australia, New Guinea and Melanesia in the current revolution in concepts of the evolution of the earth." *Search*, 1:178–189.

Carson, H. L., 1955. "The genetic characteristics of marginal populations of *Drosophila.*" *Cold Springs Harbor Symposium on Quantitative Biology*, 20:276–287.

Carson, H. L., D. E. Hardy, H. T. Spieth, and W. S. Stone, 1970. *The Evolutionary Biology of the Hawaiian Drosophilidae. Essays in Evolution and Genetics in Honor of Theodosius Dobzhansky* (suppl. to *Evolutionary Biology*), pp. 437–543.

Chaney, R. E., 1940. "Tertiary forests and continental history." *Geol. Soc. Amer. Bull.*, 51:469–488.

Chapman, G. C., and S. B. Jones, Jr., 1971. "Hybridization between *Senecio smallii* and *S. tomentosus* (Compositae) on the granitic flatrocks of the southeastern United States." *Brittonia*, 23:209–216.

Chaudhuri, C. R., and J. B. Chatterjee, 1969. "Ascorbic acid synthesis in birds: A phylogenetic trend." *Science*, 164:435–436.

Choate, H. E., 1912. "The origin and development of the binomial system of nomenclature." *The Plant World*, 15:257–263.

Clark, W. A., 1969. "Museums for living microorganisms." *Bioscience*, 19:421–424.

Clausen, J., 1951. *Stages in the Evolution of Plant Species.* Cornell University Press. Ithaca, New York.

Cohen, D. M., and R. F. Cressey, 1969. "Natural history collections – past, present, and future." *Proc. Biol. Soc. Wash.*, 82:559–762.

Colbert, E. H., 1951. *The Dinosaur Book.* McGraw-Hill Book Co., New York.

Colwell, R., and W. J. Wiebe, 1970. " 'Core' characters for use in classifying aerobic, heterotrophic bacteria by numerical taxonomy." *Bull. Ga. Acad. Sci.*, 28:165–185.

Constance, L., 1953. "The role of plant ecology in biosystematics." *Ecology*, 34:642–649.

Craig, D. A., 1969. "A taxonomic revision of New Zealand Blepharoceridae and the orgin and evolution of the Australasian Blepharoceridae (Diptera:Nematocera)." *Trans. R. Soc., Biol. Sciences*, 11:101–151.

Croizat, L., 1958. *Panbiogeography.* Vol. 1, 2a, and 2b. Published by the author, Caracas, Venezuela.

Cronquist, A., 1968. *The Evolution and Classification of Flowering plants.* Houghton Mifflin Company, Boston.

Dansereau, P. M., 1957. *Biogeographie: an Ecological Perspective.* Ronald Press Company, New York.

Darlington, P. J., Jr., 1957. *Zoogeography: the Geographical Distribution of Animals.* John Wiley and Sons, New York.

Darwin, C., 1862. *The Various Contrivances by which Orchids Are Fertilized by Insects.* 2nd. ed. 1877. John Murray, London.

de Beaufort, L. F., 1951. *Zoogeography of the Land and Inland Waters.* Sidgwick and Jackson Ltd., London.

De Candolle, A., 1884. *Origin of Cultivated Plants.* Kegan Paul and Trench, London.

De Ley, J., 1969. "Molecular data in microbial systematics." Pp. 248–268 in *Systematic Biology,* U. S. National Acad. Sci. Publ. No. 1692.

Dethier, V., 1954. "Evolution of feeding preferences in phytophagous insects." *Evolution,* 8:33–54.

Dietz, R. S., and J. C. Holden. 1970. "The breakup of Pangea." *Scientific American* , 223 (4):30–41.

Digby, L., 1912. "The cytology of *Primula Kewensis* and of other related *Primula* hybrids." *Ann. Bot.,* 26:357–388.

Dobzhansky, T., 1951. *Genetics and the Origin of Species,* 3d ed. Columbia University Press, New York.

──────1958. "Evolution at work." *Science,* 127:1091–1098.

──────1970. *Genetics of the Evolutionary Process.* Columbia University Press, New York and London.

Dobzhansky, T., and O. Pavlovsky. 1966. "Spontaneous origin of an incipient species in the *Drosophila paulistorum* complex." *Proc. National Acad. Sci.,* 55:727–733.

──────1971. "Experimentally created incipient species of *Drosophila.*" *Nature,* 230:289–292.

Dobzhansky, T., and B. Spassky, 1959. "*Drosophila paulistorum,* a cluster of species in *statu nascendi.*" *Proc. National Acad. Sci.,* 45:419–428.

Durant, W., 1939. *The Story of Civilization: Part 2. The Life of Greece.* Simon and Schuster, New York.

Edmunds, G. F., Jr., 1962. "The principles applied in determining the hierarchic level of the higher categories of Ephemeroptera." *Syst. Zool.,* 11:22–31.

Ehrlich, P. R., and P. H. Raven, 1965. "Butterflies and plants: a study in coevolution." *Evolution,* 18:586–608.

Eisner, T., R. E. Silberglied, D. Aneshansley, J. E. Carrel, and H. C. Howland, 1969. "Ultraviolet video-viewing: The television camera as an insect eye." *Science,* 166:1172–1174.

Ekman, S., 1935. *Tiergeographies des Meeres.* Akad. Verlagsges, Leipzig.

──────1953. *Zoogeography of the Sea.* Sidgwick and Jackson, London.

Elton, C. S., 1958. *The Ecology of Invasions by Animals and Plants.* Methuen and Company, Ltd., London.

Emerson, A. E., 1938. "Termite nests — A study of the phylogeny of behavior." *Ecol. Monog.,* 8:247–248.

_____ 1949. "Section V. Ecology and evolution." In Allee, et al., pp. 598–729.

_____ 1956. "Ethospecies, ethotypes, taxonomy, and evolution of *Apicotermes* and *Allognathotermes* (Isoptera, Termitidae)." *Amer. Mus. Novitates*, 1771:1–31.

Eyde, R. E., 1971. "Evolutionary morphology: Distinguishing ancestral structure from derived structure in flowering plants." *Taxon*, 20:63–73.

Fitch, W. M., and E. Margoliash, 1967. "Construction of phylogenetic trees." *Science*, 155:279–284.

Foucart, M. F., S. Bricteux-Greigoire, C. Jeuniaux, and M. Florkin, 1965. "Fossil proteins of Graptolites." *Life Sciences*, 4:467–471.

Freitag, R., 1965. "A revision of the North American species of the *Cicindela maritima* group with a study of hybridization between *Cicindela duodecimguttata* and *oregona.*" *Quaestiones Entomologicae*, 1:87–170.

Fryxell, P. A., 1957. "Mode of reproduction in higher plants." *Bot. Rev.*, 23:135–233.

Fulton, B. B., 1952. "Speciation in the tree cricket." *Evolution*, 6:283–295.

Gerald, J. W., 1971. "Sound production during courtship in six species of sunfish (Centrarchidae)." *Evolution*, 25: 75–87.

Germeraad, J. H., C. A. Hopping, and J. Muller, 1968. "Palynology of Tertiary sediments from tropical areas." *Rev. Paleobot. Palynol.*, 6:189–348.

Ghent, A. W., 1966. "The logic of experimental design in the biological sciences." *BioScience*, 16:17–22.

Ghiselin, M. T., 1972. "Book review of Provine, W. B., The origin of theoretical population genetics. University of Chicago Press, 1971." *Science*, 175:507.

Goldschmidt, R., 1940. *The Material Basis of Evolution.* Yale University Press, New Haven, Conn.

_____ 1955. *Theoretical Genetics.* University of California Press, Berkeley.

Gordon, M., 1947. "Genetics of *Platypoecilus maculatus.* IV. The sex determining mechanism in two wild populations of the Mexican platyfish." *Genetics*, 32:8–17.

Grant, V., 1848. "Pollination systems as isolating mechanisms." *Evolution*, 3:82–97.

_____ 1963. *The Origins of Adaptations.* Columbia University Press, New York.

_____ 1966. "The origin of a new species of *Gilia* in a hybridization experiment." *Genetics*, 54:1189–1199.

_____ 1971. *Plant Speciation.* Columbia University Press, New York.

Gustaffson, A., 1946–1947. "Apomixis in higher plants." *Lunds Universitets Araskrift*, 42–43:1–370.

Haartman, J., 1751, 1764. *Plantae Hybridae.* Uppsala, 1751; Amoenitates Academicae, Holm, 1764.

Haeckel, E., 1866. *Generelle Morphologie der Organismen, II.* Georg Reiner, Berlin.

Hall, D. G., 1948. "The blow flies of North America." *Thomas Say Foundation Publ.*, 4:1–477.

Hamilton, K. G. A., 1971. "A remarkable fossil homopteran from Canadian Cretaceous amber representing a new family." *Canad. Ent.*, 103:943–946.

Hammond, A. L., 1970. "Deep sea drilling: a giant step in geological research." *Science*, 170:520–521.

Handlirsch, A., 1925. "Kapitel: Die systematischen Grundbegriffe und Phylogenie und Stammesgeschichte." In Schröder, C., *Handbuck der Entomologie*, Band III. Jena.

Hanson, H. C., and R. H. Smith, 1950. "Canada geese of the Mississippi flyway, with special reference to an Illinois flock." *Bull. Ill. Nat. Hist. Surv.*, 25:59–210.

Hanson, J. F., 1960. "A case of hybridization in Plecoptera." *Bull. Brooklyn Ent. Soc.*, 55:25–34.

Hanson, N. R., 1958. *Patterns of Discovery.* Cambridge, England, by the University.

Haucke, R. L., 1971. "The effect of light quality and intensity on sexual expression in *Equisetum* gametophytes." *Amer. J. Bot.*, 58:373–377.

Heiser, C. B., 1949. "Study in the evolution of the sunflower species *Helianthus annuus* and *H. bolanderi.*" *Univ. Calif. Publ. Bot.*, 23:157–208.

———— 1951. "Hybridization in the annual sunflowers: *Helianthus annuus* × *H. debilis* var. *cucumerifolius.*" *Evolution*, 5:42–51.

Heiser, C. B., Jr., J. Soria, and D. B. Burton, 1965. "A numerical taxonomic study of *Solanum* species and hybrids." *Amer. Naturalist*, 99:471–488.

Heller, J. C., 1964. "The early history of binomial nomenclature." *Huntia*, 1:33–70.

Hennig, W., 1950. *Grundzüge einer Theorie der Phylogenetischen Systematik.* Deutscher Zentralverlag, Berlin.

———— 1966. *Phylogenetic Systematics.* Translated by D. D. Davis and R. Zangerl. Univ. Ill. Press, Urbana.

Heslop-Harrison, J., 1953. *New Concepts in Flowering Plant Taxonomy.* W. Heinemann, Ltd., London.

Hewitt, G. M., and B. John, 1972. "Inter-population chromosome polymorphism in the grasshopper *Podisma pedestris*. II. Population parameters." *Chromosoma* (Berl.), 37: 23–42.

Hilgenberg, O. C., 1933. *Vom wachsenden Erdball.* Geissmann and Bartsch, Berlin.

Hirsch, J., ed., 1967. *Behavior-genetic Analysis.* McGraw-Hill, New York.

Hoare, C. A., 1957. "The classification of trypanosomes of veterinary and medical importance." *Veterinary Rev. Annot.*, 3:1–13.

Holley, R. W., J. Apgar, G. A. Everett, J. T. Madison, M. Marquisse, S. H. Merrill, J. R. Penswick, and A. Zamir, 1965. "Structure of a ribonucleic acid." *Science*, 147:1462–1465.

Hollingshead, L., 1930. "A lethal factor in *Crepis* effective only in an interspecific hybrid." *Genetics*, 15: 114–140

Hoyle, F., and J. V. Narlikar. 1971. "On the nature of mass." *Nature*, 233:41–44.

Hsu, T. C., 1952. "Chromosomal variation and evolution in the *virilis* group of *Drosophila.*" *Univ. Tex. Publ.*, 5204:35–72.

Hubbs, C. L., 1955. "Hybridization between fish species in nature." *Syst. Zool.*, 4:1–20.

Hubbs, C. L., and L. C. Hubbs, 1932. "Apparent parthenogenesis in nature in a form of fish of hybrid origin." *Science*, 76:628–630.

Hull, D. L., 1964. "Consistency and monophyly." *Syst. Zool.*, 13:1–11.

———— 1967. "Certainty and circularity in evolutionary taxonomy." *Evolution*, 21:174–189.

_____ 1972. Personal communication.

Hunzicker, J. H., 1969. "Molecular data in plant systematics." Pp. 280–312 in *Systematic Biology*, U. S. National Acad. Sci. Publ. No. 1692.

Irving, E., 1959. "Paleomagnetic pole positions: a survey and analysis." *Geophys. J.*, 2:51–79.

Irving, L., 1953. "The naming of birds by Nunamiut Eskimo." *Arctic*, 6:35–43.

Jameson, D. L., 1955. "Evolutionary trends in the courtship and mating behavior of Salientia." *Syst. Zool.*, 4:105–119.

Jardine, N., and R. Sibson, 1971. *Mathematical Taxonomy*. John Wiley and Sons, New York.

John, B., and G. M. Hewitt, 1970. "Inter-population sex chromosome polymorphism in the grasshopper *Podisma pedestris*. I. Fundamental facts." *Chromosoma* (Berl.), 31:291–308.

Jones, S. B., 1972. "A systematic study of the Fasciculatae group of *Vernonia* (Compositae)." *Brittonia*, 24:28–45.

Jordan, D. S., 1905. "The origin of species through isolation." *Science*, n. s. 22: 545–562.

Jowett, D., 1958. "Populations of *Agrostis* species tolerant of heavy metals." *Nature*, 182: 816–817.

Kenyon, D. H., and G. Steinman. 1969. *Biochemical Predestination*. McGraw-Hill Book Company, New York.

Kernaghan, R. P., and L. Ehrman, 1970. "An electron microscopic study of the etiology of hybrid sterility in *Drosophila paulistorum*. I. Mycoplasma-like inclusions in the testes of sterile males." *Chromosoma*, 29:291–304.

Kessel, E. L., 1955. "The mating activities of balloon flies." *Syst. Zool.*, 4:96–104.

Kiauta, B., 1967. "Considerations of the evolution of the chromosome complement in Odonata." *Genetica*, 38:430–446.

_____ 1968. "Distribution of the chromosome numbers in Trichoptera in the light of phylogenetic evidence." *Genen en Phaenen*, 12:110–113.

Kinsey, A. C., 1930. *The Gall Wasp Genus Cynips, a Study in the Origin of Species*. Indiana University Studies XVI.

_____ 1936. *The Origin of Higher Categories in Cynips*. Indiana University Publ., Sci. Series No. 4.

Klikoff, L. G., 1966. "Temperature dependence of the oxidative rates of mitochondria in *Danthonia intermedia, Penstemon davidsonii* and *Sitanion hystrix*." *Nature*, 212:529–530.

Knight, H. H., 1924. "On the nature of the color patterns in Heteroptera with data on the effects produced by temperature and humidity." *Ann. Ent. Soc. Amer.*, 27:258–274.

Kolreuter, J. G., 1761–1766. "Vorläufige Nachricht von einigen des Geschlecht der Planzen betreffenden Versuchen und Beobachtugen." Reprint, 1893, in Oswald's *Klassiker*, Leipzig.

Koopman, K. F., 1950. "Natural selection for reproductive isolation between *Drosophila pseudoobscura* and *Drosophila persimilis*." *Evolution*, 4:135–148.

Kruckeberg, A. R., 1954. "Plant species in relation to serpentine soils," pp. 267–274, in "The ecology of serpentine soils: A symposium." *Ecology*, 35:258–288.

_____ 1969. "Ecological aspects of the systematics of plants." p. 161–202 in *Systematic Biology*, U. S. National Acad. Sci. Publ. No. 1692.

Kummel, B., 1970. *History of the Earth: an Introduction to Historical Geology*, 2nd ed. W. H. Freeman, San Francisco.

Lamarck, J. B., 1963. *Zoological Philosophy*, New York, Hafner Publishing Co. (Reprint of a translation of Lamarck's 'Philosophie Zoologique', 1809, Lamarck's first detailed statement of his theory.)

Lanham, W. E., 1957. "Comparative biology of the meadowlarks (*Sturnella*) in Wisconsin." *Publ. Nuttall Orinithological Club*, 1:1–67.

Latta, R., and A. Macbeath, 1956. *The Elements of Logic*. Macmillan and Company, Ltd., London.

Laven, H., 1959. "Speciation by cytoplasmic isolation in the *Culex pipiens* complex." *Cold Springs Harbor Symposia Quantitative Biology*, 20:166–173.

Lawrence, G. H. M., 1951. *Taxonomy of Vascular Plants*. The Macmillan Co., New York.

Lee, K. Y., R. Wahland, and E. Barbu, 1956. "Contenu en bases puriques et pyrimidiques des DNA des bacteries." *Ann. Inst. Pasteur*, 91:212–224.

Leonard, J. W., and F. A. Leonard, 1949. "An annotated list of Michigan Trichoptera." *Occas. Papers Mus. Zool., Univ. Mich.*, 522: 1–35.

Lewis, H., 1953a. "The mechanism of evolution in the genus Clarkia." *Evolution*, 7:1–20.

_____ 1953b. "Chromosome phylogeny and habitat preference of Clarkia." *Evolution*, 7:102–109.

_____ 1962. "Catastrophic selection as a factor in speciation." *Evolution*, 16:257–271.

_____ 1969. "Comparative cytology in systematics." Pp. 523–533 in *Systematic Biology*, U. S. National Acad. Sci. Publ. No. 1692.

Lewis, H., and M. E. Lewis, 1955. "The genus *Clarkia*." *Univ. Calif. Publ. Bot.*, 20:241–392.

Lewis, H., and M. R. Roberts, 1956. "The origin of *Clarkia lingulata.*" *Evolution*, 10:126–138.

Lewis, W. H., 1969. "Discussion." Pp. 535–538 in *Systematic Biology*, U. S. National Acad. Sci. Publ. No. 1692.

Lindroth, C. H., 1957. *The Faunal Connections between Europe and America*. Almqvist and Wiksell/Gebers, Stockholm, and John Wiley and Sons, New York.

Linsley, E. G., Jr., 1972. Personal communication.

Littlejohn, M. J., 1959. "Call differentiation in a complex of seven species of *Crinia* (Anura, Leptodactylidae)." *Evolution*, 13:452–468.

Love, A., and D. Love, 1953. "Studies on *Bryoxiphium.*" *The Bryologist*, 56: 73–94, 183–203.

Lowe, C. H., and J. W. Wright, 1966. "Evolution of parthenogenetic species of *Cnemidophorus* (whiptail lizards) in western North America." *J. Ariz. Acad. Sci*, 4:81–87.

Lowe, C. H., J. W. Wright, C. J. Cole, and R. K. Bezy, 1970a. "Natural hybridization between the teiid lizards *Cnemidophorus sonorae* (Parthenogenetic) and *Cnemidophorus tigris* (Bisexual)." *Syst. Zool.*, 19:114–127.

_____ 1970b. "Chromosomes and evolution of the species groups of *Cnemidophorus* (Reptilia: Teiidae)." *Syst. Zool.*, 19:128–141.

Lull, R. S., 1917. *Organic Evolution*. Macmillan, New York.

Mabry, T. J., 1966. "The betacyanins and betaxanthins." Pp. 231–244, in *Comparative Phytochemistry*, ed. T. Swain. Acad. Press, London.

Mabry, T. J., K. R. Markham, and M. B. Thomas, 1970. *The Systematic Identification of Flavonoids.* Springer-Verlag, New York.

MacArthur, R. H., and E. O. Wilson, 1967. *The Theory of Island Biogeography.* Princeton Univ. Press. Princeton, New Jersey.

Macmillan, C., 1959. "The role of ecotypic variation in the distribution of the central grassland of North America." *Ecol. Monog.* 29:285–308.

———— 1964. "Ecotypic differentiation within four North American prairie grasses: I. Morphological variation within transplanted community fractions." *Amer. J. Bot.,* 51:1119–1128.

———— 1965. "Ecotypic differentiation within four North American prairie grasses: II. Behavioral variation within transplanted community fractions." *Amer. J. Bot.,* 52:55–65.

Mandel, M., 1969. "Molecular data in microbial systematics." Discussion. Pp. 269–275 in *Systematic Biology,* U. S. National Acad. Sci. Publ. No. 1692.

Manton, I., 1950. *Problems of Cytology and Evolution in the Pteridophyta.* Cambridge Univ. Press.

Manwell, C. C., M. A. Baker, and W. Childers, 1967. "The genetics of hemoglobin in hybrids- I. A molecular basis for hybrid vigor." *Comparative Biochemistry and Physiology,* 10:103–120.

Maslin, T. P., 1952. "Morphological criteria of phyletic relationships." *Syst. Zool.,* 1:49–70.

———— 1968. "Taxonomic problems in parthenogenetic vertebrates." *Syst. Zool.,* 17:219–231.

———— 1971. "Conclusive evidence of parthenogenesis in three species of *Cnemidophorus* (Teiidae)." *Copeia,* 1971:156–158.

Mason, H. L., 1950. "Taxonomy, systematic botany, and biosystematics." *Madroño,* 10:193–208.

Mathad, S. B., and J. E. McFarlane, 1967. "Effect of initial rearing at 28°C followed by 35°C on wing development in *Gryllodes sigillatus* (Walk.)." *Canad. J. Zool.,* 45:135–137.

Mattfeld, J., 1930. "Über hybridogene Sippen der Tannen." *Bibliotheca Botanica,* 25:1–84.

Matthew, W. D., 1915. "Climate and evolution." *Ann. New York Acad. Sci.,* 24:171–318.

Matthews, J. V., Jr., 1970. "Two new species of *Micropeplus* from the Pliocene of western Alaska with remarks on the evolution of Micropeplinae (Coleoptera: Staphylinidae)." *Canad. J. Zool.,* 48:779–788.

Matsuda, R., 1960. "Morphology, evolution and a classification of the Gerridae (Hemiptera-Heteroptera)." *Univ. of Kansas Sci. Bull.,* 41:25–632.

Mayr, E., 1942. *Systematics and the Origin of Species.* Columbia Univ. Press, New York.

———— 1947. "Ecological factors in speciation." *Evolution,* 1:263–288.

———— 1963. *Animal Species and Evolution.* The Belknap Press, Harvard University Press, Cambridge, Mass.

———— 1969. *Principles of Systematic Zoology.* McGraw-Hill, Inc., New York.

———— 1972. "The nature of the Darwinian revolution." *Science,* 179:981–989.

McClure, J. W., and R. E. Alston, 1966. "A chemotaxonomic study of Lemnaceae." *Amer. J. Bot.*, 53:849–859.

McDaniel, I. N., and W. R. Horsfall, 1957. "Induced copulation of aedine mosquitoes." *Science*, 125:745.

McFarlane, J. E., 1964. "Factors affecting growth and wing polymorphism in *Gryllodes sigillatus* (Walk.): Dietary protein level and a possible effect of photoperiod." *Canad. J. Zool.*, 42:767–771.

_____ 1972. "Studies on vitamin E in the house cricket, *Acheta domesticus* (L.) (Orthoptera: Gryllidae). I. Nutritional albinism." *Canad. Ent.*, 104:511–514.

McMillan, C., 1954. "Parallelisms between ecology and taxonomy." *Ecology*, 35: 92–94.

Mecham, J. S., 1960. "Introgressive hybridization between two southeastern treefrogs." *Evolution*, 14:445–457.

Menzel, M. Y., and D. W. Martin, 1971. "Chromosome homology in some intercontinental hybrids in *Hibiscus* sect. *Furcaria.*" *Amer. J. Bot.*, 58:191–202.

Merrill, E. D., 1954. "The botany of Cook's voyages." *Chronica Botanica*, 14:i–iv, 161–384.

Merervey, R., 1969. "Topological inconsistency of continental drift on the present-sized earth." *Science*, 166:609–611.

Meyer, V. G., 1970. "A facultative gymnosperm from an interspecific cotton hybrid." *Science*, 169: 886–888.

Meyerhoff, A. A., 1970*a*. "Continental Drift: Implications of paleomagnetic studies, meteorology, physical oceanography, and climatology." *J. Geol.*, 78: 1–51.

_____ 1970*b*. "Continental drift, II: High latitude evaporite deposits and geologic history of Arctic and North American oceans." *J. Geol.*, 78: 406–444.

Meyerhoff, A. A., and H. A. Meyerhoff, 1972*a*. " 'The new global tectonics': Major inconsistencies." *Amer. Assoc. Pet. Geol. Bull.*, 56:269–336.

_____ 1972*b*. " 'The new global tectonics': Age of linear magnetic anomolies of ocean basins." *Amer. Assoc. Pet. Geol. Bull.*, 56:337–359.

Meyerhoff, A. A., and C. Teichert, 1971. "Continental drift, III. Late Paleozoic glacial centers, and Devonian-Eocene coal distribution." *J. Geol.*, 79: 285–321.

Michener, C. D., 1949. "Parallelisms in the evolution of the saturniid moths." *Evolution*, 3:129–141.

Michener, C. D., and R. R. Sokal, 1957. "A quantitative approach to a problem in classification." *Evolution*, 11:130–162.

Miller, A. H., 1939. "Analysis of some hybrid populations of juncos." *Condor*, 41:211–214.

_____ 1955. "A hybrid woodpecker and its significance in speciation in the genus *Dendrocopos.*" *Evolution*, 9:317–321.

Moore, B. P., and B. E. Wallbank, 1968. "Chemical composition of the defensive secretion in carabid beetles and its importance as a taxonomic character." *Proc. R. Ent. Soc. Lond.* (B):37:62–72.

Moore, H. B., 1958. *Marine Ecology*. John Wiley and Sons, New York.

Moore, R. C., and L. R. Laudon, 1943. "Evolution and classification of Paleozoic crinoids." Geol. Soc. Amer. Special Papers No. 46.

Moore, T. E., and H. H. Ross, 1957. "The Illinois species of *Macrosteles*, with an evolutionary outline of the genus (Hemiptera, Cicadellidae)." *Ann. Ent. Soc. Amer.*, 50:109–118.

Moss, W. W., 1967. "Some new analytic and graphic approaches to numerical taxonomy, with an example from the Dermanyssidae (Acari)." *Syst. Zool.*, 16:177–207.

———— 1968. "Experiments with various techniques of numerical taxonomy." *Syst. Zool.*, 17:31–47.

Müller, H. J., 1954. "Uber den Einflufs der Tageslänge auf die Saisonformenprägung von *Euscelis plebejus* Fall." *Verh. Deutsch. Zool. Ges. Tübingen:* 307–316.

———— 1958. "The taxonomic value of the male genitalia in leafhoppers in the light of new studies on the seasonal forms of *Euscelis.*" *Proc. 10th Internat. Cong. Ent.*, 1:357–362.

National Research Council, 1969. *Systematic Biology. Proceedings of an International Conference.* U. S. Nat. Acad. Sci. Publ. No. 1692.

Niethammer, J., 1969. "Zur Frage der Introgression bei den Waldmausen *Apodemus sylvaticus* und *A. flavicollis* (Mammalia, Rodentia)." *Zeit. Zool. Syst. Evolut.-Forsch.*, 7: 77–127.

Njoku, E., 1956. "Studies on the metamorphosis of leaves. II. The effect of light intensity on leaf shape in *Ipomoea caerulea.*" *New Phytol.* 55: 91–110.

Notini, G., 1941. *Om harens biologi.* Svensk. Jagarforb. Meddelande 4. Uppsala.

———— 1948. *Hararna.* Svensk. djur. Daggdjuren, Stockholm.

Ornduff, R., 1969. "The systematics of populations in plants." Pp. 104–125, in *Systematic Biology*, U. S. National Acad. Sci. Publ. No. 1692.

Parkes, K. C., 1951. "The genetics of the golden-winged × blue-winged warbler complex." *Wilson Bull.* 63:5–15.

Parsons, P. A., 1967. *The Genetic Analysis of Behavior.* Methuen, London.

Patterson, J. T., and W. S. Stone, 1952. *Evolution in the Genus Drosophila.* Macmillan, New York.

Pauling, L., and E. Zukerkandl, 1963. "Chemical paleogenetics: Molecular 'restoration studies' of extinct forms of life." *Acta Chem. Scan.*, 17 (Suppl. no. 1):9–16.

Peters, W. L., and G. F. Edmunds, Jr., 1970. "Revision of the generic classification of the eastern hemisphere Leptophlebiidae (Ephemeroptera)." *Pacific Insects*, 12:157–240.

Petersen, C. G. J., 1914. "Valuation of the sea. II. The animal communities of the sea-bottom and their importance for marine zoogeography." *Rep. Danish. Biol. Stat.*, 21:1–68.

Pimental, D., G. J. C. Smith, and J. Soans, 1967. "A population model of sympatric speciation." *Amer. Nat.*, 101:493–504.

Pitelka, F. A., 1951. "Speciation and ecologic distribution in American jays of the genus *Aphelocoma.*" *Univ. Calif. Publ. Zool.*, 50:195–464.

Platt, J. R., 1964. "Strong inference." *Science*, 146:347–353.

Popper, K. R., 1959. *The Logic of Scientific Discovery.* Hutchinson and Company, Ltd., London.

Porter, K. R., 1969. "Evolutionary status of the Rocky Mountain population of wood frogs." *Evolution*, 23:163–170.

Prakash, S., R. C. Lewontin, and J. L. Hubby, 1969. "A molecular approach to the study of genic heterozygosity in natural populations. IV. Patterns of genic variation in central, marginal, and isolated populations of *Drosophila pseudoobscura.*" *Genetics*, 61: 841–858

Price, J. L., 1958. "Cryptic speciation in the *vernalis* group of Cyclopidae." *Canad. J. Zool.*, 36:285–303.

Randall, J. H., Jr., and J. Buchler, 1942. *Philosophy, an Introduction.* Barnes and Noble, New York.

Raven, P. H., B. Berlin, and D. E. Breedlove, 1971. "The origins of taxonomy." *Science*, 174:1210–1213.

Remington, C. L., 1968. "Suture-zones of hybrid interaction between recently-joined biotas." *Evol. Biol.*, 2:321–428.

Repenning, C. A., 1967. "Palearctic-Nearctic mammalian dispersal in the late Cenozoic." Pp. 288–311 in *The Bering Land Bridge*, D. M. Hopkins, ed., Stanford Univ. Press, California.

Richards, P. A., and A. G. Richards, 1969. "Acanthae: A new type of cuticular process in the proventriculus of Mecoptera and Siphonaptera." *Zool. Jb. Anat.*, 86:158–176.

Ricker, W. E., 1952. "Systematic studies in Plecoptera." *Ind. Univ. Publ., Sci. Ser.*, 18:1–200.

Rinkel, R. C., 1965. "Microgeographic variation and covariation in *Pemphigus populi-transversus.*" *Univ. Kansas Sci. Bull.*, 46:167–200.

Roberts, H. F., 1929. *Plant Hybridization before Mendel.* Princeton Univ. Press, Princeton, New Jersey.

Rohlf, F. J., 1963. "Congruence of larval and adult classification in *Aedes* (Diptera: Culicidae)." *Syst. Zool.*, 12:97–117.

———— 1967. "Correlated characters in numerical taxonomy." *Syst. Zool.*, 16:109–126.

Romanes, G. J., 1897. *Darwin and after Darwin. Part 3.* Open Court Publishing Company, London.

Romer, A. S., 1959. *The Vertebrate Story*, 4th ed. Univ. Chicago Press, Chicago.

Ross, C. A., 1967. "Development of fusulinid (Foraminiferida) faunal realms." *J. Paleont.* 41:1341–1354.

———— 1969. "Paleoecology of *Triticites* and *Dunbariella* in Upper Pennsylvanian strata of Texas." *J. Paleont.*, 43: 298–311.

———— 1970. "Concepts in late Paleozoic correlations." *Geol. Soc. Amer., Special Paper* 124:7–36.

Ross, H. H., 1929. "Sawflies of the subfamily Dolerinae of America north of Mexico." *Ill. Biol. Monog.*, 12(3):116pp.

———— 1937. "A generic classification of the Nearctic sawflies (Hymenoptera, Symphyta)." *Ill. Biol. Monog.*, 15 (2):173pp.

———— 1953. "Polyphyletic origin of the leafhopper fauna of *Ilex decidua.*" *Trans. Ill. Acad. Sci.*, 46:186–192.

———— 1956. *Evolution and Classification of the Mountain Caddisflies.* Univ. Ill. Press, Urbana.

———— 1958. "Evidence suggesting a hybrid origin for certain leafhopper species." *Evolution*, 12:337–346.

———— 1962. *A Synthesis of Evolutionary Theory.* Prentice-Hall, Inc., Englewood Cliffs, N. J.

———— 1963. "The dunesland heritage of Illinois." *Ill. Nat. Hist. Surv., Cir.* 49: 1–28.

———— 1964. "Evolution of caddisworm cases and nets." *Amer. Zool.*, 4:209–220.

———— 1970. "The ecological history of the Great Plains: Evidence from grassland insects." Pp. 225–240 in *Pleistocene and Recent Environments of the Central Great Plains*. Dept. of Geology, Univ. of Kansas, Special Publication 3. Univ. Kansas Press, Lawrence.

———— 1972a. "The origin of species diversity in ecological communities." *Taxon*, 21:253–259.

———— 1972b. "An uncertainty principle in ecological evolution." Pp. 133–164 in *A Symposium on Ecosystematics*, Occas. Paper 4, Univ. Ark. Mus.

Ross, H. H., G. C. Decker, and H. B. Cunningham, 1965. "Adaptation of temperate phylogenetic lines from tropical ancestors in *Empoasca*." *Evolution*, 18:639–651.

Ross, H. H., and K. G. A. Hamilton, 1970. "Phylogeny and dispersal of the grassland leafhopper genus *Diplocolenus* (Homoptera: Cicadellidae)." *Ann. Ent. Soc. Amer.*, 63:328–331.

Ross, H. H., and W. E. Ricker, 1971. "The classification, evolution, and dispersal of the winter stonefly genus *Allocapnia*." *Ill. Biol. Monog.*, 45: 1–166.

Ross, J. P., 1967. "Champlainian Ectoprocta (Bryozoa), New York State." *J. Paleo.*, 41:632–648.

Rudd, R. L., 1955. "Population variation and hybridization in some California shrews." *Syst. Zool.*, 4:21–34.

Runcorn, S. K., 1959. "Rock magnetism." *Science*, 129:1002–1012.

Rutten, M. G., 1962. *The Geological Aspects of the Origin of Life on Earth.* Elsevier Publ. Co., Amsterdam.

Sailer, R. I., 1954. "Interspecific hybridization among insects with a report on cross-breeding experiments with stink bugs." *J. Econ. Ent.*, 47:377–383.

Sauer, J., 1957. "Recent migration and evolution of the dioecious amaranths." *Evolution*, 11:11–31.

Savile, D. B. O., 1956. "Known dispersal rates and migratory potentials as clues to the origin of the North American biota." *Amer. Midland Nat.*, 56:434–453.

Scandalios, J. G., 1969. "Genetic control of multiple molecular forms of enzymes in plants: A review." *Biochemical Genetics*, 3: 37–79.

Schmidt, R. S., 1955a. "The evolution of nest-building behavior in *Apicotermes* (Isoptera)." *Evolution*, 9:157–181.

———— 1955b. "Termite (*Apicotermes*) nests — important ethological material." *Behaviour*, 8:344–356.

Schuchert, C., 1924. *A Textbook of Geology. Part II: Historical Geology*, 2nd rev. ed. John Wiley and Sons, New York.

———— 1955. *Atlas of Paleogeographic Maps of North America.* John Wiley and Sons, New York.

Sclater, P. L., 1858. "On the general geographical distribution of the members of the class Aves." *J. Proc. Linn. Soc.* (London), *Zoology*, 2:130.

Seevers, C. H., 1957. "A monograph on the termitophilous Staphylinidae." *Fieldiana: Zoology*, 40:1–334.

Selander, R. B., 1964. "Sexual behavior in blister beetles (Coleoptera). I. The genus *Pyrota*." *Canad. Ent.*, 96:1037–1082.

Selander, R. B., and J. M. Mathieu, 1969. "Ecology, behavior, and adult anatomy of the Albida Group of the genus *Epicauta* (Coleoptera, Meloidae)." *Ill. Biol. Monog.*, 41: 168 pp.

Selander, R. K., 1969. "The ecological aspects of the systematics of animals." Pp. 213–247 in *Systematic Biology*, U. S. National Acad. Sci. Publ. No. 1692.

Selander, R. K., and R. F. Johnston, 1967. "Evolution in the house sparrow. I. Intrapopulation variation in North America." *Condor* 69:217–258.

Selander, R. K., M. H. Smith, S. Y. Yang, W. E. Johnson, and J. B. Gentry, 1971. "IV. Biochemical polymorphism and systematics in the genus *Peromyscus*. I. Variation in the oldfield mouse (*Peromyscus polionotus*)." *Univ. Texas Publ.* 7103: 49–90.

Shaw, C. R., and R. Prasad, 1970. "Starch gel electrophoresis of enzymes—a compilation of recipes." *Biochemical Genetics*, 4: 297–320.

Sibley, C. G., 1954. "Hybridization in the red-eyed towhees of Mexico." *Evolution*, 8:252–290.

Sibley, C. G., and D. A. West, 1958. "Hybridization in the red-eyed towhees of Mexico: the eastern plateau populations." *Condor*, 60:85–104.

Simpson, G. G., 1944. *Tempo and Mode in Evolution*. Columbia Univ. Press, New York.

——— 1952. "How many species?" *Evolution*, 6:342.

——— 1953. *The Major Features of Evolution*. Columbia University Press, New York.

——— 1961. *Principles of Animal Taxonomy*. Columbia Univ. Press, New York.

Skerman, V. B. D., 1959. *A Guide to the Identification of the Genera of Bacteria*. Williams and Wilkins, Baltimore.

Smith, D. M., and D. A. Levin, 1963. "A chromatographic study of reticulate evolution in the Appalachian *Asplenium* complex." *Amer. J. Bot.*, 50:952–958.

Smith, H. M., and F. N. White, 1956. "A case for the trinomen." *Syst. Zool.*, 5:183–190.

Smith, I., 1960. *Chromatographic and electrophoretic techniques. Vol. 2. Zone electrophoresis*. 2nd. ed. John Wiley and Sons, New York.

Smith, P. W., 1957. "An analysis of post-Wisconsin biogeography of the prairie peninsula region based on distributional phenomena among terrestrial vertebrate populations." *Ecology*, 38:205–212.

Sokal, R. R., N. N. Heryford, and J. R. L. Kispaugh, 1971. "Changes in microgeographic variation patterns of *Pemphigus populi-transversus* over a six-year span." *Evolution*, 25:584–590.

Sokal, R. R., and P. H. A. Sneath, 1963. *Principles of Numerical Taxonomy*. W. H. Freeman, San Francisco.

Sokal, R. R., and P. A. Thomas. 1965. "Geographic variation of *Pemphigus populi-transversus* in eastern North America: Stem mothers and new data on alates." *Univ. Kansas Sci. Bull.*, 46:201–252.

Solbrig, O. T., 1970. *Principles and Methods of Plant Biosystematics*. Macmillan Company, New York.

Sonneborn, R. M., 1965. "Nucleotide sequence of a gene: first complete specification." *Science*, 148:1410.

Spassky, B., R. C. Richmond, S. Perez-Salas, O. Pavlovsky, C. A. Mourão, A. S. Hunter, H. Hoenigsberg, T. Dobzhansky, and F. J. Ayala, 1971. "Geography of the sibling species related to *Drosophila willistoni*, and of the semispecies of the *Drosophila paulistorum* complex." *Evolution*, 25:129–143.

Spieth, H. T., 1952. "Mating behavior within the genus *Drosophila* (Diptera)." *Amer. Mus. Nat. Hist. Bull.*, 99:395–474.

Stalker, H. D., 1953. "Taxonomy and hybridization in the Cardini group of *Drosophila.*" *Ann. Ent. Soc. Amer.,* **46**:343–358.

Stearn, W. T., 1959. "The background of Linnaeus's contributions to the nomenclature and methods of systematic biology." *Syst. Zool.,* **8**:4–22.

Stebbins, G. L., Jr., 1942. "The genetic approach to problems of rare and endemic species." *Madroña,* **6**:241–258.

_____ 1950. *Variation and Evolution in Plants.* Columbia Univ. Press, New York.

_____ 1959. "The role of hybridization in evolution." *Proc. Amer. Phil. Soc.,* **103**:231–251.

Steere, W. C., 1937. "*Bryoxiphium norvegicum,* the sword moss, as a preglacial and interglacial relic." *Ecology,* **18**:346–358.

_____ 1938. "Critical bryophytes from the Keweenaw Peninsula, Michigan, II." *Ann. Bryologici.,* **11**:145–152.

_____ 1954. "Chromosome number and behavior in arctic mosses." *Bot. Gazette,* **116**:93–133.

_____ 1958. "Evolution and speciation in the mosses." *Amer. Nat.,* **92**:5–20.

Stone, W. S., 1962. "The dominance of natural selection and the reality of superspecies (species groups) in the evolution of *Drosophila.*" Studies in Genetics, Univ. Texas Publ. No. 6205:507–537.

Stone, W. S., W. C. Guest, and F. D. Wilson. 1960. "The evolutionary implications of the cytological polymorphism and phylogeny of the virilis group of *Drosophila.*" *Proc. U. S. Nat. Acad. Sci.* **46**:350–361.

Straw, R. M., 1955. "Hybridization, homogamy, and sympatric speciation." *Evolution,* **9**:441–444.

_____ 1956. "Floral isolation in *Penstemon.*" *Amer. Nat.,* **90**:47–53.

Strickland, H. E., 1842. "Rules for zoological nomenclature." *Brit. Assoc. Adv. Sci. Rept.,* 1842:105–121.

Sturtevant, A. H., 1939. "On the subdivision of the genus *Drosophila.*" *Proc. Nat. Acad. Sci.,* **25**:137–141.

_____ 1942. *The Classification of the Genus Drosophila, with Descriptions of Nine New Species.* Univ. Texas Publ. No. 4213, 6pp.

Sweadner, W. R., 1937. "Hybridization and the phylogeny of the genus *Platysamia.*" *Ann. Carnegie Mus.,* **25**:163–242.

Taylor, T. N., 1969. "Cycads: Evidence from the upper Pennsylvanian." *Science,* **164**:294–295.

Teichert, C., 1956. "How many fossil species?" *J. Paleon.,* **30**:967–969.

Thiel, R., 1957. *And There was Light.* (Translated from the German by Richard and Clara Winston). The New American Library, New York.

Thoday, J. M., and J. B. Gibson, 1962. "Isolation by disruptive selection." *Nature,* **193**:1164–1166.

Thompson, P. E., D. S. English, and W. Bleecker, 1969. "Genetic control of the hemoglobins of *Chironomus tentans.*" *Genetics,* **63**:183–192.

Thompson, P. E., and G. Patel, 1972. "Compensatory regulation of two closely related hemoglobin loci in *Chironomus tentans.*" *Genetics* **70**: 275–290.

Thorne, R. F., 1963. "Some problems and guiding principles of angiosperm phylogeny." *Amer. Nat.,* **97**:287–305.

Thorpe, W. H., 1961. *Bird song*. Cambridge Monog. in Exptl. Biol. No. 12.

Thouless, R. H., 1953. *Straight and Crooked Thinking*. Pan Books, Ltd., London.

Throckmorton, L. H., 1962. "X. The problem of phylogeny in the genus *Drosophila*." Pp. 208–343 in *Studies in Genetics. II. Research Reports on Drosophila Genetics, Taxonomy and Evolution*. Univ. Texas Publ. No. 6205.

———— 1968a. "Biochemistry and taxonomy." *Ann. Rev. Ent.*, 13:99–144.

———— 1968b. "Concordance and discordance of taxonomic characters in *Drosophila* classification." *Syst. Zool.*, 17:355–387.

———— 1969. "Roundtable discussion: Molecular systematics — a view of the future." Pp. 382–384 in *Systematic Biology*, U. S. National Acad. Sci. Publ. No. 1692.

Tinkle, D. W., 1969. "Evolutionary implications of comparative population studies in the lizard *Uta stansburiana*." Pp. 133–153 in *Systematic Biology*, U. S. National Acad. Sci. Publ. No. 1692.

Tralau, H., 1967. "The phytogeographic evolution of the genus *Ginkgo L.*" *Botaniska Notiser*, 120:409–422.

———— 1968. "Evolutionary trends in the genus *Ginkgo*." *Lethaia*, 1:63–101.

Turesson, G., 1922. "The genotypical response of the plant species to the habitat." *Hereditas*, 3:211–350.

———— 1925. "The plant species in relation to habitat and climate." *Hereditas*, 9:81–101.

Turner, B. L., 1967. "Plant chemosystematics and phylogeny." *Pure and Applied Chemistry*, 14:189–213.

Udine, E. J., 1941. "The black grain sawfly and the European wheat stem sawfly in the United States." U. S. D. A. Cir. 607:1–9.

Usinger, R. L., 1966. "Monograph of Cimicidae (Hemiptera-Heteroptera)." *Thomas Say Foundation*, 7:1–585.

Uzzell, T. M., Jr., 1969. "Unisexual species of salamanders." *Discovery*, 4:99–108.

Uzzell, T. M., Jr., and S. M. Goldblatt, 1967. "Serum proteins of salamanders of the *Ambystoma jeffersonianum* complex, and the origin of the triploid species of the group." *Evolution*, 21:345–354.

Van der Kloot, W. G., and C. M. Williams, 1953a. "Cocoon construction by the Cecropia silkworm: I. The role of the external environment." *Behaviour*, 5:141–157.

———— 1953b. "Cocoon construction by the Cecropia silkworm: II. The role of the internal environment." *Behaviour*, 5:157–174.

Vaurie, C., 1950. "Notes on Asiatic nuthatches and creepers." *Amer. Mus. Nov.*, 1472:1–39.

———— 1951. "Adaptive differences between two sympatric species of nuthatches." *Proc. 10th Int. Ornith. Congr.*, :163–166.

Vavilov, N. I., 1951. "The origin, variation, immunity and breeding of cultivated plants." *Chronica Botanica*, 13:i–xviii + 364. Translated by K. S. Chester.

Wagner, M., 1868. "Über die Darwinsche Theorie in bezug auf die geographische Verbreitung der Organismen." *Sitzber. bayer. Akad. Wiss. (Math-natur.)* 1868 (1) :359–395.

———— 1889. *Entstehund der Arten durch raumliche Sonderung*. Gesammelte Aufsatze. Basel. Schwabe.

Wagner, W. H., Jr., 1969a. "The construction of a classification." Pp. 67–103 in *Systematic Biology*, U. S. National Acad. Sci. Publ. No. 1692.

_____ 1969*b*. "The role and taxonomic treatment of hybrids." *BioScience* 19:785–789.

Wakeland, C., 1958. "The high plains grasshopper." *U.S.D.A., Tech. Bull.,* 1167:1–168.

Wallace, A. R., 1876. *The Geographical Distribution of Animals.* 2 vols. Macmillan, London.

Wallace, B., 1968. *Topics in Population Genetics.* W. W. Norton, New York.

Wallbank, B. E., and D. F. Waterhouse, 1970. "The defensive secretions of *Polyzosteria* and related cockroaches." *J. Insect Physiol.,* **16**: 2081–2096.

Wardle, P., 1963. "Evolution and distribution of the New Zealand flora, as affected by Quarternary climates." *N. Z. J. Bot.,* 1:1–17.

Warner, R. E., 1971. Personal communication, Dec. 20.

Wasserman, A. O., 1970. "Polyploidy in the common tree toad *Hyla versicolor* LeConte." *Science,* 167:385–386.

Waterhouse, D. F., and A. R. Gilby, 1964. "The adult scent glands and scent of nine bugs of the superfamily Coreoidae." *J. Insect Physiol.,* 10:977–987.

Wegener, A., 1912. "Die Entstehung der Kontinente," Petermanns Mitteilungen, pp. 185–195, 253–256, 305–309.

_____ 1924. *Origin of Continents and Oceans.* Methuen, London.

Wells, J. W., 1963. "Coral growth and geochronometry." Nature, 197(4871):948–950.

Wheeler, N. H., 1942. "Trap-light studies on leafhoppers of the genus *Empoasca* (Homoptera: Cicadellidae), 1932–1941." *Proc. Ent. Soc. Wash.,* 44:69–72.

Wheeler, W. M. 1907. "The polymorphism of ants, with an account of some singular abnormalities due to parasitism." *Bull. Amer. Mus. Nat. Hist.,* 23:1–94.

White, M. J. D., 1954. *Animal Cytology and Evolution,* 2nd. ed. Cambridge Univ. Press.

_____ 1956. "Adaptive chromosomal polymorphism in an Australian grasshopper." *Evolution,* 10:298–313.

_____1968. "Models of speciation." *Science,* 159:1065–1070.

White, M. J. D., H. L. Carson, and J. Cheney, 1964. "Chromosomal races in the Australian grasshopper *Moraba viatica* in a zone of geographical overlap." *Evolution,* 18:417–427.

Williamson, D. L., and L. Ehrman, 1967. "Induction of hybrid sterility in nonhybrid males of *Drosophila paulistorum.*" *Genetics,* 55:131–140.

Willis, B. 1896. "The northern Appalachians." Pp. 169–202 in *The Physiography of the United States,* Major J. W. Powell, ed. American Book Company, New York.

Wilson, E. O., 1959. "Studies of the ant fauna of Melanesia. VI. The tribe Cerapachyini." *Pacific Insects,* 1:39–57.

Wilson, E. O., and W. L. Brown, Jr., 1953. "The subspecies concept and its taxonomic application." *Syst. Zool.,* 2:97–111.

Yang, T. W., 1970. "Major chromosome races of *Larrea divaricata* in North America." *J. Ariz. Acad. Sci.,* 6:41–45.

Yen, J. H., and A. R. Barr, 1971. "New hypothesis of the cause of cytoplasmic incompatibility in *Culex pipiens* L." *Nature,* 232: 657–658.

Young, D. A., Jr., 1953. "Empoascan leafhoppers of the solana group with descriptions of two new species." *J. Agri., Univ. of Puerto Rico,* 37: 151–160.

Index